# 797,885 Books
are available to read at

www.ForgottenBooks.com

Forgotten Books' App
Available for mobile, tablet & eReader

ISBN 978-1-332-05668-2
PIBN 10277494

This book is a reproduction of an important historical work. Forgotten Books uses state-of-the-art technology to digitally reconstruct the work, preserving the original format whilst repairing imperfections present in the aged copy. In rare cases, an imperfection in the original, such as a blemish or missing page, may be replicated in our edition. We do, however, repair the vast majority of imperfections successfully; any imperfections that remain are intentionally left to preserve the state of such historical works.

Forgotten Books is a registered trademark of FB &c Ltd.
Copyright © 2015 FB &c Ltd.
FB &c Ltd, Dalton House, 60 Windsor Avenue, London, SW19 2RR.
Company number 08720141. Registered in England and Wales.

For support please visit www.forgottenbooks.com

# 1 MONTH OF FREE READING

at

www.ForgottenBooks.com

By purchasing this book you are eligible for one month membership to ForgottenBooks.com, giving you unlimited access to our entire collection of over 700,000 titles via our web site and mobile apps.

To claim your free month visit:
www.forgottenbooks.com/free277494

\* Offer is valid for 45 days from date of purchase. Terms and conditions apply.

English
Français
Deutsche
Italiano
Español
Português

# www.forgottenbooks.com

**Mythology** Photography **Fiction** Fishing Christianity **Art** Cooking Essays Buddhism Freemasonry Medicine **Biology** Music **Ancient Egypt** Evolution Carpentry Physics Dance Geology **Mathematics** Fitness Shakespeare **Folklore** Yoga Marketing **Confidence** Immortality Biographies Poetry **Psychology** Witchcraft Electronics Chemistry History **Law** Accounting **Philosophy** Anthropology Alchemy Drama Quantum Mechanics Atheism Sexual Health **Ancient History Entrepreneurship** Languages Sport Paleontology Needlework Islam **Metaphysics** Investment Archaeology Parenting Statistics Criminology **Motivational**

# THE HUMAN EAR

## ITS IDENTIFICATION AND PHYSIOGNOMY

BY

MIRIAM ANNE ELLIS

"Lend me your ears"
*Jul. Cæs.* ACT III. SC. 2

WITH ILLUSTRATIONS FROM PHOTOGRAPHS (COPYRIGHT)
CHIEFLY FROM NATURE-PRINTS

LONDON
ADAM AND CHARLES BLACK
1900

# PREFACE

THE following chapters contain a method of classifying portraits of the human ear, by which reference for the purpose of identification is made possible and convenient. They also show the value of identification among "non-criminals," by means of a minute division of the shapes of the rim of the ear.

The subject of heredity, as shown by the shape of the ear, is illustrated by nature-prints from members of several families. No one has hitherto investigated this branch of the subject.

A concise account, from original sources, of ancient and modern writers upon Ears is given, ending with a full analysis of Lavater's tentative efforts in that direction. Since his time there has been no systematic scientific investigation of the peculiarities of the forms of the ear and their important place in physiognomy, except the very recent examination of criminals' ears alone. The "non-criminal" classes have a still more distinctive and better-developed ear.

Nature-printing from the ear was invented by me, for the purpose of having permanent portraits of ears of the exact size and shape of the originals, by which means they could be compared and collated. The illustrations have been reduced uniformly in size, the proportions being kept the same.

M. A. E.

Oxford, 1900.

# CONTENTS

| CHAPTER | PAGE |
|---|---|
| 1. The Human Ear as a Means of Identification | 1 |
| 2. The Law of the Shape and Proportion of the Outer Ear | 13 |
| 3. Classification of Ears | 27 |
| 4. Aristotle and Pliny on Ears | 43 |
| 5. Mediæval Writers on Ears | 54 |
| 6. Della Porta, Ghirardelli, and Rureis | 65 |
| 7. Lavater on the Ears | 86 |
| 8. Lavater's Outlines of Ears | 104 |
| 9. The Physiognomy of the Ear | 123 |
| 10. Further Examples of the Ear | 147 |
| 11. Heredity as shown in Ears | 166 |
| 12. Ears as Portrayed in Sculpture and Painting | 180 |
| 13. Concerning Ear-rings and Ear-lore | 198 |
| 14. The Ear in Literature and Science | 210 |
| Index | 221 |

*With photographic illustrations (copyright).*

# LIST OF ILLUSTRATIONS

| FIG. | | PAGE |
|---|---|---|
| 1. | The Five Divisions | 15 |
| 2. | The Eight Squares | 17 |
| 3. | The Ear-Shell | 24 |
| 4. | Twenty-one Ears (Lavater) | 95 |
| 5. | Nine Ears (Lavater) | 106 |
| 6. | Twelve Ears (Lavater) | 113 |
| 7. | Three Ears (Lavater) | 120 |
| 8, 9. | Ears of a Child | 126 |
| 10, 11, 12, 13, 14. | Divisions (1), (2), (3), (4), (5) | 132-135 |
| 15. | Ears with deficient *helix* | 143 |
| 16. | Ears of a composer of music | 144 |
| 17. | Ears of a traveller | 147 |
| 18. | Ears of an artist | 148 |
| 19. | Ears of an artist and traveller | 149 |
| 20. | Divisions pushed out of place | 151 |
| 21. | Indent in the *helix* | 152 |
| 22. | Ears of an American | 153 |
| 23. | Ears of a Royal Navy Hospital nurse | 154 |
| 24. | Ears of a zoologist | 155 |
| 25. | Ears of a philologist | 156 |

| FIG. | | PAGE |
|---|---|---|
| 26. | Ears of the Editor of the English Dictionary | 157 |
| 27. | Divisions alike in both ears | 158 |
| 28. | Peculiar orifice | 159 |
| 29. | Ears of a scientific discoverer | 160 |
| 30. | The Five Divisions in the bulged forms | 161 |
| 31. | Ears of Charles Dickens | 163 |
| 32-35. | Parents and two sons *facing* | 166 |
| 36, 37. | Mother and son | 168 |
| 38-44. | Parents, two daughters, and three sons *facing* | 170 |
| 45-47. | Parents and daughter | 174, 175 |
| 48-51. | Parents, daughter, and son *facing* | 176 |
| 52, 53. | Twin sisters | 178 |

# THE HUMAN EAR

## CHAPTER I

### THE HUMAN EAR AS A MEANS OF IDENTIFICATION

SINCE there are two ears to each head, it would seem that ears are particularly well adapted for personal identification. A pair affords a double chance of variety, and if one of them should come to harm the other is left to keep guard over the hearing and the outward identity. That we can only see one ear at a time leads us to bestow the undivided attention on it that is the best aid to memory. As it cannot meet our gaze like the eyes, nor turn itself up at us like the nose, nor twitch about with a torrent of words like the mouth, it cannot disconcert us whilst we are gazing at it. With the air of a sentinel to guard the head it yet stands like a host to welcome us, its doors ever open.

The system of classification used for the

identification of criminals is not sufficiently varied to include all types of ear. The "non-criminal" classes have not yet received the scientific attention they deserve, and nevertheless they hold the key to a complete system of classification, for there is more variety in the shape of the ear among the educated upper classes of every country than among the lower classes. After classifying the more complicated forms it is not difficult to apply the same system to the less elaborate shapes. In the course of observation of the ear as part of the physiognomy of the face, I have noticed that certain types went with particular characteristics. The ear cannot be altered by ordinary use, and it preserves its type when the other features are changed by time or circumstance. This forms a clue to the intricate maze of the convolutions of the outer ear. After studying many thousands of British and foreign ears—black, brown, and white—from nature, as well as innumerable photographs and authentic portraits in painting, drawing, sculpture, and casts from the life, enough specimens of the various shapes have passed before me from which to formulate certain *data* for identification as well as for classification.

# MEANS OF IDENTIFICATION

We must begin by saying that there is no hard-and-fast rule as to a good or bad ear. There is, however, a very distinct law of shape and proportion of the different parts of the outer ear, to which good and bad alike have to submit. Before stating this law we must first examine the different parts of the ear.

The *pinna* is the scientific name of the whole outer ear, as distinguished from the inner portion, which contains the parts for transmitting sound to the nervous centres. It is known to vary in appearance in every individual, and almost in each one of each pair of ears. This in itself forms a basis for identification which is invaluable. These unchanging marks of identity remain, though every actor learns how to make himself look like somebody else with very little help from paint or dyes; indeed, there have not been wanting some who were able to disguise the features by the use of the facial muscles alone. The nose can be shortened or lengthened about a quarter of an inch either way. The eyebrows make the forehead smaller or larger according to their elevation. The eyes can be altered in apparent size by the position of the eyelids, and the cheeks and mouth can attain to a still greater change of shape and size by their numerous muscles, as every schoolboy knows. Even the chin can be brought forward or pulled inward at pleasure. When we consider how very much alike people of the same age and class endeavour to make themselves by force of custom, dress, and education, we can understand how easily a slight difference could be imitated and an identification obtained on false pretences. In other days great store was set by moles or birth-marks as being proofs of identity, but moles are now known to alter in shape and size, and sometimes to disappear. Warts are often mistaken for moles, and warts have a power of vanishing as if by magic, being in reality only a nervous complaint of the skin.

The finger-tip identification, advanced by Mr. Francis Galton, shows itself to be of singular value on careful examination. It is unfortunately difficult to obtain good impressions, and photography would require life-size portraits and extreme accuracy. This difficulty appears to be very slight in theory; in practice, however, it is found to interfere

# MEANS OF IDENTIFICATION 5

greatly with the gathering together of many specimens among the "non-criminals," who are not compelled to register their finger-tips.

The want of some quick and easy way of identification without trouble to the observer or observed has often been felt. As we have said, there is but one feature that cannot be altered at will, and that is the ear. Any attempt at disguising it can be detected, and of course it cannot be pulled out of shape and place, although here and there ears are met with that can be slightly moved by the muscles from the base in one whole piece. Piercing for ear-rings cannot be obliterated, as Xenophon testifies when he speaks of the ears of Apollonides, which were "pierced after the Lydian manner," by which the unlucky man was discovered to be a Lydian, and was at once turned out of the ranks of the famous Ten Thousand (*Anab.* Bk. III. ch. i. § 31, ἐπεὶ ἐγὼ αὐτὸν εἶδον, ὥσπερ Λυδὸν, ἀμφότερα τὰ ὦτα τετρυπημένον, lit. *But I see him like a Lydian, having been pierced as to both his ears*). This is a very early case of identification by the ear!

On account of the advantages of keeping the hair short for men, or tied up for women, the ear has been very generally shown; and its beauty helps to give attraction to the side-face. It can be photographed with ordinary care, and ear-prints can be obtained direct from nature. Moreover, the ear fulfils the condition of being recognisable without trouble or inconvenience to the observer or observed.

The ears alter slightly in shape during childhood, until they attain their full growth. This requires a passing reference, because it has been taken for granted that the ear does not alter in shape from its first appearance in the world. M. Alphonse Bertillon says: " Les variations de confirmation si nombreuses que présente cet organ paraissent subsister sans modification depuis la naissance jusqu'à la mort" (*La Photographie Judiciaire*, Paris 1890). We must submit that the assumption came from want of *data*, criminals not being found willing to be photographed, nor likely to be done in early childhood. By reference to photographs of children, so frequently taken from a few months old onwards in the families of the educated upper

# MEANS OF IDENTIFICATION

classes ("non-criminals") the slight alteration can be perceived. Figs. 8 and 9 show earprints. taken from a child at different ages (see chap. ix.). The ear undergoes enlargement in conformity with the increasing size of the nose and jaw, altering in shape to tally with them. The *helix*, or outer rim, has been known to curl over after birth. The exact age at which the ear is full grown differs with the individual, but it would appear to finish developing before the full height of the figure is attained. Growing girls and boys are often noted for their big ears amongst their parents and guardians, and yet when the nose and jaw have grown to adult size the ears are declared to be "not so big" as formerly. They cannot have shrunk, but they have sunk into their right proportion to the rest of the face and are no longer noticeably large. After childhood the ears do not alter till late in life, when they enlarge slightly without losing their shape. A good photograph of the full-grown ear will therefore form a permanent record by which the owner can be identified at any period.

Although the fact that the adult ear does

not alter like the other features has been known and used for the identification of criminals for some years, its value in the case of the "non-criminal" classes has been overlooked. If every young man left life-size photographs of his ears behind him when he started on his travels round the world, or in search of military glory, or of colonial success, it would be impossible for pretenders to take his place if he were missing. History might have been materially altered, and romances would have lost a valuable aid for their enlivenment, if it had been known long ago that the nature-print of the ear could calmly and certainly dispose of false claimants to thrones, to properties, or to the heroine's affections. The difficulty of recognising twins is overcome by this simple means, for they each have their special ear in spite of constant similarity of the other features. A man could also be found, if alive, with absolute certainty, for there is no way of disguising the ears without showing the traces of the attempt, and the pair being seldom a good match there is a double chance of personal identification. Even if one ear were lost or injured, it is very un-

# MEANS OF IDENTIFICATION 9

likely that both should be destroyed. Moreover, a man who has lost his ears by any misfortune is generally known to have done so, and this would attract that attention to him which might lead to his eventual recognition in a way that would not happen unless the search had been conducted on these lines.

From careful observation of many thousands of ears I have found that the *pinna*, or outer ear, varies much in size and in shape, the right ear being usually the largest, whilst the *helix*, or outer rim of the ear, varies even more in shape. The *tragus*, or little projection from the cheek towards the lower part of the hollow of the ear, varies both in shape and position. The *anti-tragus*, or scoop of the lower part of the hollow of the ear (often having a little projection in it), is placed either near the *tragus* or far off it, in an extraordinary variety of shapes and positions. The *lobe*, or small pendant at the end of the *helix* at the bottom of the ear, differs as constantly from its fellow as from others. The *top* of the ear is usually rounded, though occasionally it is straight, and sometimes it is sloped upwards almost into a peak. Some ears possess a little *knot* in the edge of the rim. This is the knot described by Darwin. It is so uncommon that it becomes of particular use for personal identification, rarely being in the same part of each ear, and seldom in both the ears of the same person. Sometimes this knot appears at the edge of the ear when there is no rim at all.

The branching undulations that fill in the upper part of the ear above the hollow up to the *helix* are called the *anti-helix*. It has not been found necessary to use them in the identification of "non-criminals," where the other varieties of shape are so numerous. They are said to be of material assistance in the identification of criminals, where, the points of identification being fewer, every possible item has to be employed. Their shape forms a kind of ribbing to support the flap of the *pinna*, and probably affects the refraction of the waves of sound. The *pinna*, or outer ear, is adapted to the catching of the waves of sound, which get shot backwards and forwards into the hollow opening and then along the inner tube up to the drum of the ear. It is not the size of the *pinna*, but its shape, which is calculated to transmit the waves of sound with the least loss of strength. The hearing further depends upon the make and condition of the inner ear and upon the perfection of the auditory nerves. The connection between the inner and the outer ear will be considered later on.

*To take a nature-print of an ear.*—Place some printing-ink on a piece of glass; take a roller and roll out the ink till the roller as well as the glass is smoothly covered with it. Then apply the roller to the ear. The *helix* and *lobe*, and parts of the *tragus* and *anti-tragus*, will be blackened. Place a piece of fine-grained printing-paper upon the ear; hold it firmly with the left hand, pressing it down gently and evenly with the right hand. On removing it the nature-print will be found. The outline of the *helix* and the outline of the *hollow* of the ear must be accentuated in pencil and afterwards traced in ink, to allow of reproduction by photography. The process must be done quickly, as the ink soon dries. The ink must be removed from the ears by a soft rag dipped in turpentine or in oil of cloves, after which the ears must be washed with soap and hot water and well dried with a soft towel.

It is best to use the glass, glue-roller, and specially prepared printing-ink, and strips of paper six inches long by two inches wide, that Mr. Francis Galton has had prepared for taking finger-prints. They can be obtained from 357 Oxford Street, W.

# CHAPTER II

## THE LAW OF THE SHAPE AND PROPORTION OF THE OUTER EAR

It will be convenient first to describe the divisions of the *helix*, which are necessary for the purposes of classification. They are five in number, and their position at a first glance might appear to be arbitrary, especially in the bulged form adopted in Fig. 1. The reason why they are not likely to be recognised at a moment's inspection, is because it is very unusual to see all five at once in the same ear. When they are carried in the mind, or sketched out for reference, as in Fig. 1, all varieties will be seen to be formed by the presence or absence of any one of these divisions. If two or more divisions are run together into one distinct curve, this shape is ranged among the *coalescing* forms that make subdivisions for classification and identification.

It was not until after I had found that Five Divisions of the *helix*, exclusive of the lobe, were required for the classification of the ears of "non-criminals," that I learned that the French system of *four* divisions, exclusive of the lobe, was still in use in France, and was employed in England. The French *first* division consists of about a quarter of an inch of the *helix* where it roots itself in the hollow of the ear. Their *second* division begins where I count Division (1) to begin, and they carry it right over to the other side of the top of the ear, thus including what I call Division (1) and Division (2) in one sweep without distinction. Their *third* division includes the part I call Division (3), together with a piece of Division (4); whilst their *fourth* division takes the rest of what I call Division (4) and also Division (5). On examining photographs of English criminals, through the kindness of the Head of the Identification Office in Scotland Yard, it could be seen that the French system continued to hold good for criminals of another nation, in a marked absence of the part I call Division (1) and a very rare appearance of the place where I

## SHAPE AND PROPORTION

put Division (2). The "non-criminal" classes appear to have more fully developed ears than the criminal, since a larger number of divisions is required to classify them. The converse nevertheless is not true, *i.e.* all ears with a simple *helix* do not indicate criminals,

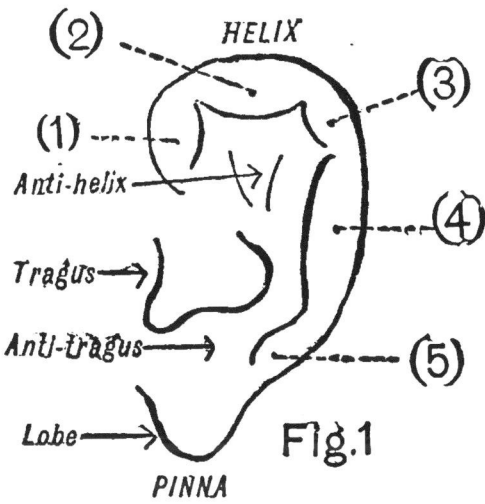

Fig. 1

although criminals have a comparatively simple *helix*.

The Five Divisions of the Helix (Fig. 1)—
(1) Is often short and generally straight.
(2) Is about twice the length of (1).
(3) Is the shortest of all.
(4) Is as long or longer than (2).
(5) Is about the same length as (2).

More than Five Divisions have not been found, except as a modification of some one or other division. The *helix* may be jagged or knotted, so as to distract the beginner with the fear that divisions may lurk for him without end, though the practised observer will find that the jags and knots keep in their divisions with an irregular regularity and belong to them. Sometimes the *helix* will have a deep perpendicular indent in Division (4), though hardly ever in the other divisions. At times the rim is absent in whole or in part, as if it had been snipped off if the ear is narrow, or as if it had been flattened out if the ear is wide. The *pinna*, or whole outer ear, is flat to the head in the accepted canons of beauty. Every kind of sticking-out or curving over assists in identification.

The law of the shape and proportion of the different parts of the outer ear may now be stated. The length of the *pinna* from the top of the ear to the bottom of the lobe will be found to be the same length as the nose, measured from the top, where it joins the forehead exactly between the eyebrows, to the bottom, where the inner wall dividing the nostrils meets the upper lip. This is

# SHAPE AND PROPORTION

the proportion known to artists; it is founded upon the natural growth.

The width of the *pinna* should be at its middle part exactly half its length, and this should also be the widest part. Any deviation from these exact measurements at once forms a valuable aid in identification.

A piece of glass will now be required with eight squares drawn upon it with ink, or ruled with a diamond point. This should be laid upon an ear-print, and the curves of the *helix* and the proportion of the *pinna* will fall into place as it were, in a manner to make it possible to compare varieties with definiteness of aim (see Fig. 2).

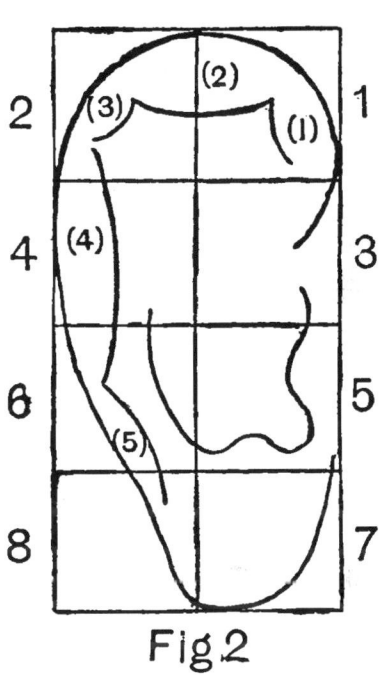
Fig 2.

The four upper squares will hold, as in a frame, the curves of the *helix* at (1), (2), (3), and part of (4). The four lower squares will contain the rest of (4) and all (5), and the lobe, and also the *tragus* and the *anti-tragus*. A medium-sized oval ear should be chosen from which to make the measurements. As a rule, men's ears are about a quarter of an inch longer and broader than a woman's. On placing the squares on any ear-print and examining where the parts fit in, the proportion of the Five Divisions of the *helix* and the way in which they run into one another can be perceived. We first observe how much of each square is taken up by the *helix*, and whether the top of the ear is straight or curved, and how far it may chance to be curved in proportion to the whole length. The way in which the ear slopes off at Division (3) will catch the eye, and also how far back the *pinna* goes in that direction, and whether or not it has a square corner at that part. The size of the *hollow* of the ear will then be distinctly apparent. This is a part of the ear that is rarely drawn correctly; it is generally made too small or too narrow, or it is squeezed into the part next to the cheek, in order to make the ear appear unnaturally small. This is done even in woodcuts from photographs, unless extreme accuracy is aimed at for scientific purposes.

## SHAPE AND PROPORTION  19

By the above simple expedient, its size and proportion to the whole ear is readily seen and easily drawn. As the *pinna* is shaped for the purpose of reflecting waves of sound into its orifice, the shape and the size of the hollow of the ear are of distinct importance. It is as necessary for fidelity of identification that the opening should not be drawn in the wrong part of the ear, as that in sketching a face the mouth should not be drawn in the middle of the nose. We could not properly recognise a face so disarranged, nor can we fully identify an ear when the hollow is wrongly placed in the drawing.

The use of a medium-sized ear by which to draw the squares, will also serve to show how much larger or smaller exceptional ears may be. Our eyes are so little trained to gauge the size of the ear that no one can be sure of it at a first glance, and on account of the accepted contraction of sketched ears, the tell-tale ear-print is hailed as impossibly large. Very small ears have been considered a sign of good family, but this is indeed a baseless superstition. Men and women alike have ears that match their noses, and whenever they pride themselves on having the large nose, also considered a sign of good family, they must be content to take the desired smallness of their ears on trust, for a moment of measurement will dispel the illusion. That small ears are to be found among members of old families is due to those members not happening to have extra large noses, and not to the fact of their pedigree. Ears that are small and beautifully modelled are found in every rank of life. They are always in company with short noses, measured in the way described. If the tip of the nose should slant downwards over the lip, the nose will appear to be larger than the above proportion, and this will cause the ear to look smaller; but by careful measurement it will always be found that the nose is short from the top between the eyebrows to the joining at the upper lip, and is of the same length as the small ear. Where the ear appears to be enormous in comparison to the nose, the same measurement will prove this to be an illusion, depending again on the shape of the tip of the nose, which often in such a case takes a turn upwards that disguises the true

## SHAPE AND PROPORTION

length of the nose from the forehead to the upper lip. Where there is a real difference, this is caused by more or less deformity of the *pinna* or of the nose, and the identification is still more completely simplified. It must also be remembered that the ear, whilst retaining the same shape, enlarges slightly in later life, and in that case exceeds the proportion allotted.

There is generally a gentle slope backwards of the whole ear in its position on the side of the head, shown by drawing a line through the centre of the top of the *pinna* to the centre of the lobe. This, as a rule, is seen to be parallel to the slope of the nose, from its top where it joins the forehead to its tip in the air, when the nose is fully grown—a process, by the way, that takes about twenty-five years. Later on the nose alters very much, but as the ear remains the same, the original slope of the nose can always be taken from it. This helps much in identifying early photographs of the same face. The same piece of glass with upright squares can be used with sloping ears; it is only necessary to place the top of the centre line upon the middle of the top of the ear, and the bottom of the centre line upon the middle of the lobe. All the rest of the ear then falls into place as before. In taking ear-prints, the slip of paper —six inches long by two inches wide—is placed slanting along a slanting ear, so that the ear-print appears to be upright when finished, and can be still more easily measured on the square-marked glass.

The outline of the *helix* or rim, from Division (4) down along Division (5), follows the shape of the outline of the jaw, while the size of the lobe tallies in proportion with the size of the chin as seen in profile. This concurrence of form is so remarkably persistent that any discrepancy of outline at once points to deformity either of the *pinna* or of the jawbone. It is not easy to note this where the cheek and chin are plump and well covered, or where a full beard is worn. This part of the outline of the ear therefore gives useful information as to identification where a beard has been added either by nature or by art for purposes of disguise.

The upper half of the ear is sometimes almost square from the flatness of the top

## SHAPE AND PROPORTION

and the extension of the *pinna* backwards. Sometimes also the lower half of the ear is almost square, from the widening of the hollow and the "setting-back" of the *antitragus* further away from the cheek. These square forms are common enough one at a time, more often in the lower half, but it is so rare to find both the upper and the lower half of an ear squared in this way, that for mere identification the combination is invaluable.

The perfect form of ear with respect to art and beauty would seem to be of an oval shape, gently curved at the top and sloping off below to a well-rounded lobe, with a delicately-curved rim embodying Divisions (2), (4), and (5), which should all be slightly bulged at the inner edge and tapered at each end. This is the type that artists and sculptors choose in their happiest moments of selection. It belongs both to men and women, and it is of the kind that poets rave about as "shell-like," though anything more unlike a real shell cannot be imagined. There is, however, a tiny, flat, and very yellow-brown shell actually called the "ear-shell" or "sea-ear," from a certain monstrous resemblance it possesses to the human ear (Fig. 3).

There is even a trace of the Five Divisions of the *helix*. Division (2) runs right into Division (5), and the lobe appears to be turned backwards up on to the top of the place where the *antitragus* should be. This is not an ear to be coveted as a beauty! The depression and the obverted lobe are white, and there is of course no hollow nor any interior at all (*Auris marina*, a genus of shell-fish. It is a section of Gasteropods, *Haliotidæ*, called "ear-shells" or "sea-ears." See Chambers's *Cyclopædia* and the *Penny Cyclopædia*). The perfect form of ear with regard to science, is shown by its possession of all the Five Divisions. All five must be well developed and distinct, and the slope of the ear at Division (4) should not be great, whilst Division (5) should be rather smaller and narrower than Division (4). There should be a tendency to squareness in the upper half,

Fig.3

## SHAPE AND PROPORTION

and the hollow of the ear should be well developed. This form is peculiarly rare and is seldom repeated in its fellow ear. It is not often seen in pictures or in sculptures, unless the heads are avowed and accurate portraits. Division (3) is so often absent that even when it is in the real ear, great artistic conscientiousness is required to carve it or to paint it, as it seems to enlarge the ear too much to please the eye of the general public. To the educated eye of the otomorphologist this full form will give a scientific satisfaction, though not always an artistic delight. The shape belongs both to men and to women, both to large and to small ears. It is as rare as complete forms of every sort must be, and when an ear has been dowered thus, there is a very remarkable clue to identification.

Much of these two chapters was given in a short paper by the present writer, which was read before the Anthropological Section of the British Association, September 8, 1898, at the Bristol meeting. By tables taken from a certain number of selected ears belonging to the educated upper classes, including notabilities in art, science, and literature, it was shown that the upper part of the left ear and the lower part of the right ear were chiefly distinctive for purposes of identification. The converse here appears to hold, viz. that when the top of the left ear and the bottom of the right ear are in no way remarkable, the owner is not likely to be a notability in art, science, or literature. Why this should happen is beyond the scope of this work and deals rather with the science of biology than with that of anthropology. It does not follow that they may not be noted in some other walk of life. In this we are anticipating the purely physiognomical part of the study of the ear, which must be considered later on.

PERCENTAGES OF DIVISIONS OF THE HELIX IN 100 EARS.

(*As between the ears.*)

|  | Absent. | Thick. | Bulged. |
| --- | --- | --- | --- |
| Division (1) . | 2 more | 5 more | 7 less |
| Division (2) . | 8 less | 8 more | equal |
| Division (3) . | 8 less | 4 less | 12 more |
| Division (4) . | 7 less | 1 more | 6 more |
| Division (5) . | 1 less | 1 less | 2 more |

*More* or *less* in the right ear than in the left ear.

# CHAPTER III

## CLASSIFICATION OF EARS

CLASSIFICATION of ears becomes possible when they are sorted under the Five Divisions of the *helix*, both singly and taken in groups of divisions. The piece of glass with eight squares drawn on it must be laid over the nature-print or life-size photograph.

The ears should first be sorted by size into three heaps :—I. Large; II. Medium; III. Small.

Again applying the squares to these three heaps in succession, we must sort each heap separately under the Five Divisions and their combined forms in the following order:—

The Five Divisions (separate and combined) :—

(1); (2); (3); (4); (5);
(1, 2); (1, 2, 3); (2, 3); (2, 3, 4); (2, 3, 4, 5);
(3, 4); (3, 4, 5); (4, 5); (1, 2, 3, 4, 5).

This table presents fourteen headings, of which very few will occur oftener than one per cent.

Each of these headings requires a further subdivision into five forms:—

    I. Bulged.
    II. Thick, or long.
    III. Very small, or thin.
    IV. Absent.
    V. Peculiar.

In using this classification we first look to see which division is the most bulged, because the bulging form is the most distinctive of the Five Divisions of the *helix*. Should none be bulged, we take the longest division for reference. Perhaps they are all distinct and all the same length; in that case the ear will go under the subdivision "Peculiar" of the last heading. It is not likely that any ear will be found of this nature, though once in the world it may have existed.

If two divisions are bulged, the ear will go under one of the headings above, according to which two divisions have coalesced, and the same method is used for noting down three or four bulged divisions. When all Five Divisions

## CLASSIFICATION OF EARS 29

run into one another without a break, it is caused by a solidly thick rim all the way round the ear, and then the rim is generally of an even thickness and not bulged. This ear would go under the subdivision "Thick," of the fourteenth, or last, heading. More common, though not frequent, are the four divisions running together from the middle of the top of the ear down to the lobe. In that case Division (1) will usually be small, and Division (2) will be pulled partly out of place towards Division (3), making the top of the ear somewhat peaked where Division (2) begins. This kind of ear is generally thick all along the rim from the peak at the top, and in that case it would go under the subdivision "Thick" of the heading (2, 3, 4, 5); but should any part of the *helix* be bulged enough to show itself apart, the whole ear must be put under the subdivision "Bulged" of the tenth heading, *e.g.* Divisions (tenth heading), Subdivision 1 (4). Here (4) means that the *helix* is bulged at Division (4).

Perhaps Division (2) and Division (3) are run together, and also Division (4) and Division (5). Observe which coalition is the largest, and place the ear under that one. Should any one of the Five Divisions be distinctly marked and bulged, it claims the whole ear, even if there are one or two others run together and also bulged. I have arranged the fourteen headings in their order of precedence, finding by experience that this particular arrangement answers best for purposes of classification and of reference for identification.

It would be a great advantage if right and left ear-prints were kept for reference in separate cabinets. They differ so greatly that they cannot be kept in pairs, and must in any case go under different headings and sub-divisions.

There are some other important marks for identification, but they are not in sufficient numbers to form a basis of classification like the *helix*. Such, for instance, is the shape of the *pinna* at the top, side, and lobe. These shapes are partly accounted for under the preliminary heaps of Large, Medium, and Small. For large ears are generally long, medium ears are often oval, and small ears are somewhat round. Squared tops are seldom found except amongst large ears. The squared

## CLASSIFICATION OF EARS

lower half makes a large ear of what would otherwise be of medium size, or it turns a small ear into a medium ear. Really small ears are nearly always rounded off or are very narrow. It is extraordinarily rare to find a small ear that has the square shape both at top and below, and indeed it is very unusual to find this kind of square form both above and below in any ear, even in the large ones

The place of the *anti-tragus* chiefly causes the shape of the lower half of the ear. It varies so much in outline that it is used as a basis of classification under the French system, but it requires an expert in constant practice to measure it quickly by the eye, even sufficiently to compare it with a photograph. Nothing short of a photograph can give it accurately, though ear-prints approximate fairly well to its shape where the parts are firm enough to impress themselves on the paper. The hollow which it outlines naturally confuses the eye in computing the size, and shape, and position of the *anti-tragus* in respect to the *tragus*, when judged from the living ear.

Another valuable point is the *knot* in the

rim. Sometimes it occurs two or three times in the same rim in different divisions or in the same division. It should be indicated on the ear-print by an arrow.

Occasionally an *indent* is formed in the face of the *helix*, as already mentioned, or on the back of the rim. These indents should be noted on the margin of the ear-print.

When ears stick very much out or forward it is difficult to take a nature-print, and the fact must be mentioned to explain the peculiar thickened or foreshortened appearance of the ear-print.

The advantage of ear-prints is the ease with which they can be collected, collated, and preserved. Being done with printing-ink they resemble finely-drawn woodcuts, and their relative sizes can be compared at a glance. Classification becomes at last possible, and the ear can be described in writing or telegraphy in reference to each square and the parts of the ear that appear in each. For this reason I have numbered the eight squares in pairs (see Fig. 2, chap. ii. p. 17). *Square* 1 is always where Division (1) should be found. *Square* 2 is the second top one, and all the uneven

# CLASSIFICATION OF EARS

numbers come below *Square* 1, and all the even numbers below *Square* 2. The squares are thus numbered so as to read from right to left, or from left to right, according to which ear is being described. By this method the even numbers will always hold the chief part of the *helix*. It is necessary to mention whether an ear-print or a photograph is being described, as the ear-print is reversed, like a negative; and whilst looking like a portrait of the right ear, as in a photograph, is really a representation, reversed, of the left ear. Consequently the squares will read from right to left for a *left* ear-print or for a *right* ear-photograph, and from left to right for a *right* ear-print or for a *left* ear-photograph.

For instance, Fig. 2, if regarded as a nature-print of the *left ear*, would read from right to left, and would be described in words as follows :—

### LEFT EAR-PRINT

*Sq.* 1. Div. (1). Subdiv. I.—Half of Div. (2). Subdiv. I.

*Sq.* 2. Half of Div. (2). Subdiv. I.—

Div. (3). Subdiv. I.—Top of Div. (4). Subdiv. I.

*Sq.* 3. Top of wide *concha*, and base of Div. (1).

*Sq.* 4. Div. (4), except top and end, Subdiv. I.

*Sq.* 5. *Tragus*, deep narrow *inlet, point* of *anti-tragus.*

*Sq.* 6. End of Div. (4). Subdiv. I.—Top of Div. (5). Subdiv. I.—*Anti-tragus* nearly level, scooped back squarely.

*Sq.* 7. *Lobe* long, thick, rounded, three-quarters.

*Sq.* 8. End of Div. (5). Subdiv. I.— Quarter of *lobe.*

From this description the ear could be sketched in numbered squares whilst it was being read out. The description can be telegraphed, telephoned, or printed. An eye trained like that of the artist, the doctor, or the physiognomist is required to judge of the ear with reference to its classification as well as to its identification. For as in the other features of the face an acute observer will at once note differences in their placing in different faces, so in the same way the Five

## CLASSIFICATION OF EARS 35

Divisions are fitted into the *helix* differently in each person. A few notes as to the appearance and probable varieties of form may be of use :—

Division (1) is often of so slightly defined a form that it counts as "absent." In this case it does not mean strictly the same as in the others of the Five Divisions, for being the turn of the ear directly after it springs out of the *concha* and cheek there must always be some rudimentary rising. But the *helix* frequently does not begin there with any definite folding over, or, at most, with a very slight and slender one. To have Division (1) "bulged" is not common either, and the "knot" is rarely found there. When Division (1) runs into Division (2) it gives a thick ending of the top of the ear next the cheek. This will tilt up the top of the ear slightly in the prick-eared way for which Lavater was at a loss to account, as we shall see later on. If Division (1) runs into Division (2), and Division (2) also runs into Division (3), the result is a thick straight line along the top of the ear, from the neck of the ear where it joins the upper part of the cheek. This is a very unusual form, and it can only be detected by applying the squares, by which the line will be found to extend too far beyond the position of the end of Division (2) in *Square* 2 to be counted as anything but as ending in Division (3), which is also in *Square* 2.

Owing to the positions of the places of Division (4) and Division (5) in the *helix* there can be no further running together of the upper and side divisions, unless the ear is totally deformed, or unless the whole rim is one compact, thick, curved mass, from beginning to end.

When Division (2) and Division (3) are run together this gives a thickness at the outer curve at the top of the ear, tilting the ear upwards towards the end of the top next the cheek. This form must not be confused with what is caused by a large kind of Division (3). By laying the squares on the ear-print we can see whether the form is Divisions (2, 3) or only Division (3). In the latter case it means that Division (3) is unusually large, and it will be found that Division (2) has in consequence been pushed out of its place in the centre of the top and

## CLASSIFICATION OF EARS 37

lies over towards the place where Division (1) should end. At the same time Division (1) is usually "absent" altogether.

We find that Division (3) is very often "absent," and it is indistinguishable in form—though not in place—when Division (2) and Division (4) are run together, when it can only be known as a connecting centre by the fact that there is no indent at all to hint at a separation between Division (2) and Division (4). For when Division (3) is really "absent," that is, when Division (2) and Division (4) are run together without any tapering, there is a slight nick or a tiny indent to show where Division (2) ends and Division (4) begins. This is sometimes the shape of the *helix*, when there is a tendency to squareness of the upper half of the ear,—a shape of ear admired by Lavater and specially desired by the early otomorphologists. It is noticeable that it is the absence of one of the Five Divisions that gives this coveted form.

Another difficult form to analyse is when Division (2) is present in its full force, and a curve at the bend of the ear in *Square* 2, down into *Square* 4, seems to indicate that

Division (3) is run into Division (4). It is often found, on examining the shape of Division (4), that this first appearance is deceptive, and that Division (3) is really "absent," but that Division (4) has been pushed a little too high in the *helix* in order to leave room for a longer tapering than usual towards Division (5). We may compare this displacement of Division (4) to the "high" cheek-bones in a face, which are not called any other name, although they are not in the usual place. In the same way we may call this Division (4) "high."

Division (4) is the longest in a normal ear. Experience shows, however, that a normal ear is as uncommon as statistics usually prove a normal form of anything to be. And Division (4) is often shortened to make room for a large Division (5); or it is pushed upwards to allow a small ear to be delicately formed, with tapering ends of Division (4) both upwards towards Division (2) and downwards towards Division (5). These tapering ends add much to the beauty of an ear, but they take up room and push the divisions a little out of place. Division (1) and Division (3) are often "absent" in this kind of ear. One very peculiar fact is that Division (4) is often larger in the right ear than in the left, so that the left ear is the favourite one for portraying, because it is more delicately shaped.

Usually the running together of divisions takes place with the group of Divisions (2, 3, 4), beginning where the middle of Division (2) should be, and ending where the middle of Division (4) should be, shortening both of them in the process and absorbing Division (3) to strengthen the union. Sometimes Division (2) and Division (3) are folded over together in their entire conjoined length, leaving but little space for Division (4), which then, being pushed out of place, puts forth its division half-length, or even quarter-length, as if afraid to take up too much space. When Division (4) is long without being bulged, it gives a firm, straight outline to the *pinna* and a tendency to leave room for squareness of the ear, either in the upper or lower half, or in both halves. Lavater has drawn the *helix* several times with this straight form.

Division (5) is the lowest; it gets lost towards the lobe, and it is frequently flattened out there. It is often found to run up so high that it pushes Division (4) out of the way altogether, and tapers off above the middle of the ear. This tapering will show that Division (5) has not absorbed Division (4), for when Division (4) and Division (5) run together the tapering does not appear till beyond Division (4), high up near the top of the ear. This running together of Division (4) and Division (5) gives what is called a "heavy" ear, a shape much adorned by ear rings, for which the lobe gains support by the strength of Division (5).

As already observed, the *anti-tragus*, when well set back, allows the orifice to be large, and thus the ear is squared at the base without any perceptible size of Division (5). On the other hand, when the squareness is formed by the large size of Division (5), the orifice is often narrow, and the *anti-tragus* is curved in a sudden slant upwards.

The *lobe* varies in size, as we have seen, according to the size of the chin in profile. There are all kinds of superstitions—ancient, yet still in existence—respecting the position of the lobe of the ear, whether it is detached from the cheek or apparently sewn on without any scrap to spare. The superstitions are wont to be entirely contradictory, and there is no clue for choosing which version to adopt. They must have been passed on by hearsay, and twisted inside out in the process, and their only value is the proof they afford that the lobe has always varied in shape *in* the same manner as at the present day.

The little *knot*, which is called "Darwin's nodule," is occasionally visible, and sometimes only detected by the touch, as lying under the softer folds of the *helix*. It is very unusual, indeed, and seems to run in families.

The jagged edge of the *helix*, found chiefly in Division (4), is generally formed by the presence of one or even two knots, and, being very uncommon, it is therefore useful for identification.

If any ear is snipped, or injured, or malformed, these facts should be added to the description, but the ear should still be placed under the main divisions of the *helix*. For accidents can be mended, as can also a slight malformation; but if the division affected is mentioned, the attempted disguise in the mending will be at once detected.

The cases where next to no *helix* is seen will have to be ranged under the subdivision "absent." They must be tested by the eight squares on the glass, to find which divisions, or whether all five, are "absent." The normal ear, as seen through the squares, should be laid beside the ear to be described. The effect of the comparison is at first rather startling. Each part of the new ear to be compared seems to show of its own accord that it has or has not any or all of the Five Divisions of the *helix*. This is what is meant by the phrase, "the divisions fall into place."

A simpler and less scientific method of comparing the divisions is to take a tracing of either Fig. 1 or Fig. 2, and lay it upon the ear-print, noting where they differ.

As otology is a medical term for the science of the ear, we should prefer to use the new word (suggested by Dr. R. Garnett) *otomorphology*, the science of the shape of the ear.

# CHAPTER IV

### ARISTOTLE AND PLINY ON EARS

FROM the earliest times those philosophers, who had a practical turn of mind, set to work to catalogue the characteristics of man, and to try and fit in the outward manifestations with the inward qualities. They found it as difficult to accomplish as any other study of nature. As usual, they called in the aid of the stars, and with the help of astrology they drew wonderful pictures of faces and hands, not such as existed, but such as they imagined ought to exist. The ears alone refused to be brought into line with these convenient, if impossible, maps of character. Neither the sun, moon, nor planets turned the faintest light upon the subject. One has to gather from this omission that the astrologers must have felt, when they faced the stars and tried to fathom their mysteries by the searching eye (not wholly unassisted by the rudimentary telescope), that the ears were useless in the quest. Therefore the ears were supposed to receive no notice from the heavenly bodies in return for keeping out of their way.

Not to be baffled, these ardent and often very learned men found a new method of accounting for the shapes of ears. They compared human ears with the ears of animals, and showed great ingenuity in deducing character in conformity with the different shapes.

At the head of all, little as we should expect it, we find the name of Aristotle. He is often referred to by later writers, who appear to have accepted the work on *Physiognomy* (Ἀριστοτέλους φυσιογνωμονικά) with complete faith. It is now classed among the "suppositious" works. Whoever wrote it, the interest attaching to it is that the book was believed to contain the greatest wisdom on the subject as known either at that period or when it was forged, for the date is uncertain. In its few pages there are only a few lines relating to ears. Of animals, those of the panther are mentioned as being round rather than flat.

Furthermore, Aristotle is made to say: "They, having the ears small, are ape-like, but they, having them big, are ass-like. Any one will see at any rate that the best of dogs have moderate ears" (Φυσιογνωμονικά, VI. Οἱ τὰ ὦτα μίκρα ἔχοντες, πιθηκώθους· οἱ δὲ μέγαλα, ὀνώδεις. ἴδοι δ' ἄν τίς κε τῶν κυνῶν τοῖς ἀρίστοις μέτρια ὦτα ἔχοντες). This extract is from a volume at the Taylorian Library, Oxford.

In another still more ancient but fragmentary copy of the same book at the British Museum the following Latin version appears —"*De Auribus.* Aures eminentes et valde magne, stoliditates, garrulitatem et impruden cias signant. Aures vero parve malignitatis sunt indicia. Aures nimium rotunde homines pusillanimum signant ındocilem et dolosum. Aures jacentes et capiti adherentes pigriciam signant." ("*Concerning Ears.* Ears sticking out and extremely large indicate stupidity, garrulity, and imprudence (rashness). Ears but indeed small are tokens of malice. Ears too much rounded in shape show men little of soul, ignorant, and crafty. Ears lying close to the head mean idleness.")

The treatise on Chiromancy that went

under Aristotle's name is now known to be forged, nor does the treatise on Physiognomy bear internal evidence of the wisdom we are accustomed to associate with his name.

With Pliny we get to the first attempt to be scientific, as well as popular, on the subject of ears. In his *Natural History* (see Bohn's translation of Pliny's *Natural History*), containing twenty thousand "matters of importance," he begins by giving instances of remarkable acuteness of hearing. The noise of the Battle of Sybaris, for instance, was heard at Olympia. Cicero corrects this statement into the Battle of the River Sagras, when a hundred and thirty thousand men were defeated by ten thousand. Whether this disproportion made the noise so great, and which side shouted the loudest, we are not told. But that cannon can be heard two hundred miles away is stated in Evelyn's *Diary*, when he says he heard the guns when our fleet engaged the Dutch at that distance from England. He was in his garden at Say's Court, Deptford.

Pliny delights in the marvellous, and he next refers to "animals which hear without ears or apertures" (X. 50). He then proceeds to mix what he considers to be fact with graceful philosophising: "Man is the only animal the ears of which are immovable. It is from the natural flaccidity of the ear that the surname of Flaccus is derived. There is no part of the body that creates a more enormous expense for our women in the pearls which are suspended from them. In the East, too, it is thought highly becoming for the men even to wear gold rings in their ears."

In considering the parts of the human body to which certain religious ideas are attached (Bk. XI. ch. ciii.), he says: "The seat of the memory lies in the lower part of the ear, which we touch when we summon a witness to depose upon memory to an arrest." The goddess of Retribution, Nemesis herself, has at last found a home, as the following passage shows:— "The seat, too, of Nemesis lies behind the right ear, a goddess which has never yet found a Latin name—no, not in the Capitol even. It is to this part that we apply the finger next the little finger, after touching the mouth with it, when we silently ask pardon of the gods for having let slip an indiscreet word " Perhaps the modern form of boxing the ear has its origin in a forgotten attempt to propitiate the pagan divinities.

Aristotle, owing to the "supposititious" work attributed to him, is not let off without blame by the painstaking Pliny, who observes with a quaint dignity, but without any thought of critical doubt of the authorship of this ridiculous fragment, "I am greatly surprised that Aristotle has not only believed, but has even committed to writing, that there are in the human body certain prognostics of the duration of life. Although I am quite convinced of the utter futility of these remarks, and am of opinion that they ought not to be published without hesitation, for fear lest each person might be anxiously looking out for these prognostics in his own person, I shall still make some slight mention of the subject, seeing that so learned a man as Aristotle did not treat it with contempt." He inserts the list, of which the only point of interest in these pages is that it ends with the words "and large ears." He remarks that Trogus, whom he considers one of the very gravest of the Roman authors, says: "Largeness of the

ears is a sign of loquacity and foolishness." These latter words we have already seen in Aristotle's supposititious work on Physiognomy. Although Pliny's sagacity made him find fault with what he believed to be Aristotle's words, he evidently did not wholly disbelieve them, and he had already accepted even more odd assortments of so-called knowledge. For instance, he chronicles the following remarkable piece of natural history (Bk. VI. ch. xxxv.):—
"There are the Sesambri also" (near Ethiopia on the Nile), "a people among whom all the quadrupeds are without ears, the very elephants even." Somebody must have been making fun of him. As a balance to this information, there are the men in India, who "have ears so large as to cover their whole body" (Bk. VII. ch. ii.).

We get upon modern ground, and feel that Pliny might be living in the present days, when we meet with the following superstition connected with the ear (Bk. XXVIII. ch. v.):—
"And then, besides, it is a notion universally received that absent persons have warning that others are speaking of them by the tingling of the ears." He does not make any

difference between the ears, which is done in the modern alliterative and poetic version· "Left for love, and right for spite;" this is contradicted by the prose form· "Right for well-speaking, left for reverse," which leaves us exactly where Pliny left us. Modern science unromantically attributes tingling of the ears to action of the liver.

Amongst remedies for diseases, those for affections of the ear were possibly the most far-fetched, and perhaps, fortunately, were difficult to obtain at any time. Pliny devotes great attention to the matter. Here are a few of the recipes :—

"Pains and diseases of the ears are cured by using . . . the gall of a wild boar, swine, or ox, mixed with castor oil and oil of roses in equal proportions. But the best remedy of all is bull's gall, warmed with leek-juice, or with honey, if there is any suppuration" (Bk. XXVIII. ch. xlviii.).

He does not discriminate between the different kinds of pain or disease; it is as much as he can do to refer to deafness as an occasional symptom.

"Some persons are of opinion that it is a

# ARISTOTLE AND PLINY

good plan to wash the ears with (another) preparation in cases where the hearing is affected; while others again, after washing the ears with warm water, insert a mixture composed of the old slough of a serpent and vinegar, wrapped up in a dossil of wool. In cases, however, where the deafness is very considerable, gall warmed in a pomegranate rind with myrrh and rue, is injected into the ears; sometimes also fat bacon is used for this purpose, . . . in all cases, however, the ingredients should be warmed . . . care being taken to warm the ears before the application, and all the remedies being wrapped in wool," except of course the injections.

Pliny seems fascinated by the subject, and pursues the remedies with more zeal than discretion. Apparently some doctors were afraid of applying a cure to the diseased ear itself, for the following curious mixture is to be injected into the *other* ear, the one that is not diseased.

Centipedes "boiled with leek-juice in a pomegranate rind—it is highly efficacious, they say, for pains in the ears; oil of roses being added to the preparation, and the mixture injected into the ear opposite to the one affected."

Possibly an excellent recipe for producing counter-irritation, but also giving the patient a good chance of losing the hearing of both his ears.

Edible snails "with a small, broad shell are employed with honey as a liniment for fractured ears." Furthermore, "the thick pulp of a spider's body, mixed with oil of roses, is also used for the ears; or else the pulp applied by itself with saffron or in wool; a cricket, too, is dug up with some of its earth and applied" (Bk. XXIX. ch. xxx.).

Pliny does not altogether approve of all the recipes, especially when frogs, edible snails, pulped spiders' bodies, and beetles are used. He thinks some beetle remedies are "quite disgusting, even to hear of," with which we agree. "Yet they are used," he adds, naïvely, "such unlimited power has the medical art to prescribe as a remedy whatever it thinks fit." Gentler remedies, suggestive rather of magic than of medicine, close his exhaustive treatment of the subject. Such are the following:—

"Honey, too, in which bees have died, is remarkably useful for affections of the ears," or "wearing a cricket attached to the body as an amulet."

(Bk. XXXII. ch. xxv.): "Wool, too, that has been dyed with the juice of the murex, employed by itself, is highly useful for this purpose; some persons, however, moisten it with vinegar and nitre." "The fat of frogs, injected into the ears, instantly removes all pains in these organs. The juice of river-crabs, kneaded up with barley-meal, is a most effectual remedy for wounds in the ears."

No wonder that votive offerings in the shape of ears for recovery from deafness are not often found among Roman remains. One at the Bristol Museum, and two in the Pitt Rivers collection at the University Museum of Oxford, have come down to us, excellently modelled in terra-cotta. Otherwise we could hardly have believed that any ear, even one belonging to the conquering race of the world, could have been able to survive the medical treatment of that period.

# CHAPTER V

## MEDIÆVAL WRITERS ON EARS

THE industrious writers in former times drew up long tables of shapes of ears and the qualities they imagined went with them. After comparing these tables with one another, we are forced to believe that they indicate rather the character of the writers than that of any one else. Some lean to clemency, others to severity. The qualities are mixed up and are frequently contradictory, and where the writers are found to agree, it is unfortunately not from personal observation but from following one another's treatises. The satisfaction of being sure that a treatise was invaluable often fell to the lot of the mediæval writers, and it did not upset their conviction when they were forced to amend any statement. The lesser writers evidently copied originals and emendations straight on, amassing contradictions as if they were heaping up treasure. Still the early writers have their use in showing that ears were at least as varied in shape and of probably the same general shapes as at the present day. This is indicated more by their epithets than by their drawings; but we have no accurate shape drawn, nor measurement of size given.

In this chapter will be found the names of the principal early writers on ears, with the exception of the three most important, who will require to be treated at full length in a separate chapter later on. For the most part they wrote on Physiognomy, and included a few lines on the subject of ears. They were frequently theologians and men of science, and were evidently careful observers as far as their powers went, but they could not photograph of course, nor even draw correctly, and they were hampered by the credulous trust in the virtues of "similarity." This took root like a weed amongst them, and overran the science of the Middle Ages. Some physicians thought that anything that outwardly resembled a disease would kill the patient, others, that it would cure him. Physiognomists were convinced that if they could detect a likeness to any animal in any part of a man's face, that man must possess the qualities of that animal. They seemed to think each kind of animal was a bundle of well-defined qualities, suitable for parcelling out amongst men, as shown by the philosophers, and not disdained by the poets.

> Fertur Prometheus addere principi
> Limo coactus particulam undique
> Desectam, et insani leonis
> Vim stomacho apposuisse nostro.
> Hor. *Carm.* I. xvi.

Nowadays animals are supposed to be like men, which is the reverse of the old superstition.

Porphyry, the philosopher, pupil of Plotinus (d. 306 A.D.), is quoted by the mediæval authorities.

Averroës, the Arabian physician and philosopher (d. 1198). He translated Aristotle from the Greek into Arabic before the Jews made their version. He is less known among his own nation than among Christians.

The celebrated Albertus Magnus (1193-

1280), who was one of the greatest of the scholastic philosophers and theologians of the Middle Ages, became a Dominican and Archbishop of Ratisbon, but shortly resigned his see. He was about the most learned man of the time. Among other things he was said to have searched for the philosopher's stone, and to be a great magician. He was formerly credited with having made a man of copper, fashioning the parts at various times under certain stars. This took him thirty years. When finished, "it had the ingenuity to reveal to the said Albertus the solution of all his principal difficulties." It sounds like an allegory of thirty years' study, combined with a delight in mechanical contrivances which may have been a "recreation" to him. Thomas Aquinas, a scholar of Albertus Magnus, is reported to have broken the figure as "he could not endure its too great prattling." A machine that persisted in solving another man's difficulties must have been a wearisome companion! But the whole pretty little fable was finally discredited. Albertus Magnus taught divinity and philosophy in Germany and France. Pope Alexander IV. ordered him to visit Rome, where he became master of the sacred palace, and combined his duties with lectures on divinity. Albertus Magnus was a very little man, and the Pope thought he was still kneeling when he had got up again. After returning to Germany he retired to his cell at Cologne, and was then ordered by the Pope to preach the Crusade through all Germany and Bohemia. In 1274 he was the Emperor's ambassador at the Council of Lyons, and again retreated to Cologne, until his death at the age of eighty-seven.

Pietro d'Abano (1250-1316). He was an Italian physician and also an alchemist. He is generally referred to as "Conciliator," the name being a short reference to the full title of one of his books, *Conciliator Differentiarum Philosophorum et Medicorum*. He studied Greek, medicine, and mathematics, and went to Paris. Recalled to Padua, he was made professor of medicine, and became so celebrated as a doctor that many fables were told of him. Being enthusiastic about astrology, he caused over four hundred astrological figures to be painted on the ceiling of the public hall at Padua. These were destroyed by fire the next century, and were repainted by Giotto. "Conciliator" was twice accused of heresy, in a somewhat contradictory manner, for "some upbraided him for not believing in demons, while others attributed his extraordinary knowledge to seven familiar spirits which he kept in a bottle." He was saved from penalties by his friends the first time, and the second time he was accused he died before he was condemned. The Inquisition wished to burn his dead body, but the Paduans persuaded them to do this in effigy instead. He wrote many books, of which a dozen were often reprinted, including works on Physiognomy, Astrology, Fevers, etc. He is sometimes called Pietrus de Padua.

Bartolomeo Cocles, who lived in the fifteenth century, wrote on occult arts both under his own name and the pseudonym of Andrea Corvo, which caused much confusion, as the names were long supposed to refer to two different people. He was renowned for his knowledge of physiognomy, and he also studied grammar, medicine, surgery, mathematics, astrology, and chiromancy, a frequent course with the learned men of that period.

Jean Taisnier, a Belgian (d. about 1562). He had a great love of travel and of learning. After entering orders he became tutor to the pages of Charles V., with whom he went on the expedition to Tunis. He travelled much in Europe, and even in Asia, teaching in the academies of Rome, Ferrara, Bologna, Padua, and Palermo. Finally he retired to Cologne, where he served in the chapel of the archbishop. He was extremely proud and vain of his knowledge and of his travels, but he was not above taking to plagiarism wholesale on the subjects of astrology, physiognomy, etc., by which he obtained a great reputation, since refuted by comparison of his writings to the extant works of others.

The well-known theologian Indagine (flourished 1522) was a German who had Latinised his name Johannes von Hagen. Of good descent, he lived in his youth at various courts, and though an ecclesiastic, he was not a monk. He obtained a living near Hanau, where he lived for nearly fifty years. Having been sent as envoy to the Pope, he wrote against the evils of the Church as he saw them at Rome. In spite of his learning and piety he was given to astrology, which was rampant in Italy, France, and Germany in the sixteenth century. He also wrote on physiognomy, and gave his portrait in his book.

Jean Belot (end of sixteenth century) devoted himself to books of occult sciences. He was *curé* of Mil Monts, and wrote in French on his favourite subjects, which included physiognomy. The peculiar turn of his mind is shown by his book, in which he tries to show that " by the aid of some prayers composed of magic words," one could learn all sciences. He does not seem to have risen into high distinction, nor to have been as learned as the others, in spite of his " magic words."

Honoratus Niquetius or Nicquet, a Frenchman (1585-1667), was a Jesuit professor of rhetoric and philosophy. At one time he was Censor of Books at Rome. Then he became head of several Jesuit colleges in France, and he established a society for aiding the sick and poor at Rouen. He wrote lives of saints, and a book on Physiognomy, in which he quotes other writers. Nicquet thought *aures*, ears, came from *haurio*, I drink, because we drink the words with our ears (haurimus enim auribus voces). Whether this was genuine philology or not, Nicquet must have known Horace's original use of the same idea—

sed magis<br>
Pugnas et exactos tyrannos<br>
Densum umeris bibit aure volgus.<br>
Hor. *Carm.* II. xiii. 30.

Nicquet has a weakness for what he considers to be reasons for his facts, both reasons and facts being somewhat after the following fashion:—"Those who have small ears are clever, the reason being that the ears lack size owing to excess of heat and dryness, from which subtle spirits are made." He accepts the ancient *dictum* that large ears mean long life, even quoting Pliny as an authority—who himself had blamed Aristotle for this supposed opinion—and going a little further on in the mazes of absurdity by giving the reasons. "This means there is the best proportion between heat and damp, whence longevity arises. Some with great ears are very melancholy, for it means coldness and dryness, and this brings timidity, like asses and hares and long-eared animals. But they have a good memory. . . . This must be understood as relating to retaining well what has been taken in with difficulty, for what is once marked on what is dry, is with difficulty removed, and dryness of the brain is indicated by large ears." He is also certain that large ears are quick at hearing, "because they gather in so much air." Again, he refers to prick-eared animals being quick to hear.

Pomponius Gauricus (1482-1528), wrote on the human figure in relation to the art of sculpture, hence he took an interest in the shape of ears. He probably went to old authorities, but he does not agree with Niquetius as to size, for he says: "The square ear, of moderate size, is especially quick in hearing." He is the only man who has a kind word for projecting ears, and says they show " docility, gentleness, benevolence."

Giovanni Ingegneri, Bishop of Capo d'Istria (fl. 1615), said large ears give loquacity and long life, "because they show the blood is thick with very earthy spirits and vapours in it." This "reason" is about as valuable as all the others. Ears with the lobe detached are said to show the owner was born at night, but the contrary is asserted by other writers, probably equally well informed, so that the average reader is left precisely as ignorant as before he read the two contraries. This is not surprising, as of course the kind of lobe has nothing whatever to do with the natal hour.

# CHAPTER VI

### DELLA PORTA, GHIRARDELLI, AND RUBEIS

Two or three of the physiognomists of that period took a passion for their work, and gave the ears rather more attention than the others had done. Or perhaps by that time the curious mass of ancient lore had become heaped more high, and they shovelled it up to the last cinder. Their names were della Porta, Ghirardelli, and Rubeis.

By far the most entertaining writer is Giovanni Battista della Porta, sometimes shortened into Battista Porta (1538-1615). He was a Neapolitan physicist of good family. He wrote *Magia Naturalis,* a highly esteemed collection of physical principles and experiments, including his own discoveries. The magic-lantern appears to own him for inventor, for which countless myriads of children owe him grateful thanks. His treatise on the *Physiognomy of Man,* in six books, was printed at Venice in 1644, in an Italian translation. The second book refers to ears. It is illustrated by woodcuts of animals' heads, together with those of men believed by della Porta to resemble them. He quotes Aristotle, Pliny, Galen, Polemon, Adamantio, Conciliator, Losso, Suetonius, Columella, and Meletius; and he dares to differ from them when he feels inclined, but the reasons given do not seem to be of any more value than the reasons overthrown by them—much like dried pease shot amongst tin soldiers. For when ears, whether large or small, project too much, della Porta thinks it means timidity, " like, for example, the hare, called the long-eared, rabbits, and so on ; whilst the lion, dog, and other courageous animals have short square ears." After this, it is not to be wondered at that he considers Polemon wrong in attributing folly to dogs, "which no other philosophers do" Nevertheless, he is not wholly faithful to dogs, for he considers " long and narrow ears mean jealousy, because they are shaped like the ears of the domestic dog," having just said that dogs have short square ears!

There are, in fact, many kinds of dogs named by the philosophers very confusedly, and compared by them to men. First, they say such "dog-eared" men are generous and magnanimous, then that they are envious and hurtful; now stupid, now wise, now lovers of wild beasts. Many dogs they call vigorous and magnanimous like lions; others are wise, as dogs of the chase. Some are swift, like greyhounds, others are the guardians of the flock and ever barking, while others again guard houses and are called "wheedlers" or "flatterers." Of these last, another writer we shall consider later on (Ghirardelli) says they are by nature easily given to anger and easily flattered, they like good eating and are called greedy. These dogs are faithful and devoted to their masters, which may be supposed to infer that the "dog-eared" man will show the same qualities towards those men who are placed over them.

We may here mention that moderate-sized ears on the whole are most approved of by all the early writers. As far as we can ascertain, this selection of a favourite size does not rest upon any firmer foundation than the words of one man, Suetonius, who said Augustus "had moderate ears, and was adorned with the best of manners and the highest gifts of mind." Such a happy mean, by the law of the "ogive," could only occur once, and no man can hope to have such moderate ears as these again. Augustus therefore remains unique, and justifies to the utmost the superb praises of Horace, who did not know what an opportunity he missed by not hymning the ears of the emperor.

Della Porta has a pleasant candour in his style of writing, and a systematic way of quoting his authorities that is agreeable if unconvincing. His portrait in his book shows a personable man in the handsome dress of the period. The somewhat retreating hair leaves a forehead well rounded; arched eyebrows, large eyes, and rather hollow cheeks are completed by a small pointed beard, allowing the lips to be seen. He wears a small ruff to his jerkin and a Medici collar to his coat.

Cornelio Ghirardelli (early seventeenth century), of Bologna, to whom we lately referred, was a Franciscan. He studied

# DELLA PORTA AND OTHERS

physiognomy in the usual way by annotating previous writers. He added art and poetry to the shaky science, and made his book *Cefalogia fisonomica* the more amusing for the mixture. He gives quaint woodcuts illustrating the text, and under each monstrous presentment he places an Italian sonnet and a Latin distich. We require every aid to make out what he means, and occasionally we are rewarded by finding a passage recording observations of his own. We subjoin his ten examples of ears:—

1. *The large thick ear*

Simplex est, crasse nutritus et immemor ille,
Cui plusquam debet, crassior auris inest.

(Simple, rudely brought up, and forgetful, the more so the thicker the ear.)

Ghirardelli says he has observed men with such ears "have a long head, lips lowered, legs thick, voice harsh, slow, cold, contemners of fatigue and injuries." He does not credit large thick ears with a good memory, and in these few lines he sketches the typical country bumpkin.

In the small picture given, the ear is drawn

with more care than usual, but the orifice is unnaturally small, and the lobe is a little lump added in a perfunctory manner. The *helix* is rather narrow:—Division (1) is absent; Division (2) is rounded at the top, making it thicker; Division (3) is absent. The ear is slightly squared where Division (4) begins and goes straight down, squaring again where Division (4) joins Division (5). It is really an uncommon ear, neither rude nor unintelligent, although the small orifice and lobe are deformities, apparently drawn thus to make the rest of the ear look larger.

## 2. *The small ear*

Auribus exiguis memor est, animosus, honestus,
Pacificus, meruens, obsequiisque datus.

(With tiny ears he possesses a good memory, is courageous, honest, peaceful, deserving, and given to obedience.)

Ghirardelli nevertheless objects to these small ears and declares they show extreme malice. He believes "their smallness shows overheat, and they are quick to all ill, not valiant, but blamable and perverse."

In the woodcut this ear appears to be a

shrivelled copy of the large ear, whilst keeping the orifice and the lobe the same attenuated size, and altering the top by pointing it a little by sloping Division (2) into Division (3). It suggests the idea that the artist had but one model to draw from, which was probably his own ear, doubly reflected by mirrors.

### 3. *The long and narrow ear*

Turgidus invidia, ferus, improbus, atque malignus,
Longis, et strictis auribus esse solet.

(Swollen with envy, wild, dishonest, and malignant, he with long and narrow ears is accustomed to be.)

Ghirardelli considers this kind to belong to men who are envious and ambitious, slanderous, astute, giddy, jealous, and open to corruption. This is evidently a gleaning of bad qualities from other authors, shown by its incoherent and composite character.

Again, the ear appears to be drawn from the same model as No. 1, only the top is pushed up into a peak by elongating Division (1), making Division (2) very short, stretching Division (3) down into Division (4), which again runs into Division (5). The *helix* is narrow, and there are the same orifice and lobe as before, which are big enough for so narrow an ear.

### 4. *The large wide ear*

Auribus est longis aliquantum, stultus et amplis,
  Est vafer, est multæ garrulitatis homo.

(The man with somewhat large and wide ears is foolish, cunning, and very loquacious.)

The physiognomists do not often mention this sort of ear. Ghirardelli makes a great effort to analyse his authorities, and solemnly declares he does not think it possible for such ears to mean both malice and madness, "for madness raves, whilst malice knows its own end, hence they are incompatible." Nor does he deem them covetous of gain. At last he floats on two opinions, like a twin-steamer, joined by a "perhaps." "The great width and size of the ears will convince us of the truth of this" (*i.e.* that this shows malice), "*perhaps* even inclined to madness, as reason is no longer taken for guide."

In the illustration the artist enlarges the orifice and lobe, and boldly draws an ear that is almost circular at the top, with a narrow even rim all round. It is of a very strange shape, slightly squared at the lower part.

### 5. *The small wide ear*

> Aure refers simiam, simios imitabere mores,
> Non tibi par furris, improba quæque subis.

(With this ear thou dost resemble a monkey, and dost imitate monkey manners; thou art not equal to thieves, but thou dost undertake some bad things.)

Ghirardelli does not object as much to these ears as his predecessors do. He thinks the evil qualities of a monkey can only be attributed to a man when the whole face as well as the ear is like a monkey's. He says: "When the ears alone are like a monkey's, they only show a tendency to mock. Also they doat on animals, especially quadrupeds and even monkeys."

The artist has drawn a small round ear, with a small lobe and a rather thick rim all round. The ear is wide, but the orifice is only one half its width. Although the divisions are all run together, Division (1) is thick, and Division (5) is rather bulged. As the width of the human ear is entirely due in the lower part to a widening of the orifice which the *pinna* extends itself to hold room for, this ear is a monster of an ear, impossible for that of man, and the *helix* and lobe make it impossible for that of a monkey.

### 6. *Good ears*

Cujus habet grandes, et quadras occiput aures,
Hic forti, et magna caliditate vir est.

(This man is strong and of great warmth, whose head has large square ears.)

All the authorities quoted by Ghirardelli concur in praising this ear. He himself says the men with these ears are "valiant, astute, and of good intellect, and desirous of the ill of others, *though* they have good manners." This little touch of surprise shows that the good men Ghirardelli had met must have had the best of manners and had desired the well-being of others. He enlarges further on the subject. "They have good intellects, but are desirous of ill through severity which is far from love and affection and nature, opposed to humanity and allied with ferocity.

# DELLA PORTA AND OTHERS

They might be like Sylla and others, who delighted in wounds and the misfortunes of others."

The picture represents an ear rather square at the top, and suddenly rounded outwards from the middle, with an orifice about two-thirds its proper size and with a moderate lobe. There is a narrow rim the same size all round, except a thickening in Division (1), and a slight bulging at the bottom of Division (5).

### 7. *The uncarved round ear*

Si quem forte vides aures gestare rotundas,<br>
Ancipitem, timidum, dixeris, atque rudem.

(If perchance you see any one bearing round ears, you should term him undecided, timid, and rough.)

Ghirardelli seems to mean by "uncarved" that the *helix* is lacking in parts. He says: "As they are wanting in the right form, so the man is wanting in goodness and justice. The better carved, the better the manners; those not well carved show depraved men, rough, indocile, timid, and unreasonable, indiscreet, and importuning. The peasant often has it; he does not know the end of things, nor how to distinguish the beginning from the middle, governed only by his natural appetite. He with such ears is *indocile*, because he is generally incapable of knowledge or discipline, and unreasonable because he is never sustained by duty, but by a wayward obstinacy, and without any restrictions of good breeding, remains ever of his own opinion. Besides, he is timid in everything and can never resolve, he appears to have paralysis of the brain and to tremble beyond measure, even though there is no cause for fear." Ghirardelli then gives a short account from his own observation, having known "men with such ears who were dishonest in reasoning, full of vain laughter, presumptuous in glance, and annoying in conversation. They are tall, but have a small mouth and face, a long neck, troubled eyes, trembling eyelids, legs not very big, hands very long, the voice delicate and sonorous."

**A**fterwards Ghirardelli gives one of his customary effete reasons. "Round ears mean too much water in the composition, as women have. It causes weakness, timidity, baseness,

# DELLA PORTA AND OTHERS 77

and irresolution." We feel quite at sea with so much water!

The artist has drawn a little ear. It does not seem exactly round, being elongated, with a hanging lobe, but the top of the ear is round and the *helix* is drawn very small in parts. In fact, here is an example of the tapering ends of the divisions of the *helix*. Division (1) is small and runs into Division (2), which is tapered at the other end. Division (3) is very tiny. Division (4) is long, and bulges a little, and is tapered at each end. Division (5) is moderate, tapered at the top towards Division (4). The orifice is represented as being extremely narrow. Such an ear is not likely to be found on the typical peasant or rough man. But the artist has evidently been endeavouring to follow the writer's description of "uncarved" by tapering parts of the *helix*, though not removing them altogether. The curious fact remains that a really natural ear has been drawn—except as to the orifice—of an exceptionally well-developed type of modern times. If we could hope that it was a portrait of any actual peasant's ear of the sixteenth century, we might understand why so many talented men sprang from the lower classes in those days. For priests and monks were often recruited from amongst them, and this kind of ear is susceptible to learning.

### 8. *The moderate-sized ear*

Aurici mediocre notant virtutis amantes,<br>
Intrepidos, dociles, ingenuosque viros.

(Moderate-eared men signify those that love virtue, and are brave, docile, and candid.)

Ghirardelli fairly surpasses himself in admiration of such ears. He confirms all the good things said of them by previous writers — whether in "supposititious" works or not—he is wrapped into an ecstasy of joy in their contemplation. "Such ears are ever open to the most sane announcements. Such men gently and politely listen to all that relates to good, and hate and abhor what is ill, and take the moderate way in every place. They are on the very throne of virtue."

The illustration is not good. The orifice and lobe are rather small, and again the *helix* is the same thickness all round, and almost all the five divisions are of equal length.

# DELLA PORTA AND OTHERS 79

The ear is very slightly curved at the top. It is somewhat squared both above and below. This form is very unusual in modern times, and the head is generally well formed, to which such a squared ear belongs. But the divisions are too similar in length for natural resemblance.

### 9. *Pointed ears*

Aures explicitæ et parvæ canis instar habebant
Ingenii stolidi, stultitiæque notam.

(Ears small and extended are like unto a dog's, and are the sign of stupidity and folly)

Ghirardelli is not as hard on these ears as his favourite authors show themselves. "Dogs with little extended ears often leap and bay forth without any reason, as if they were mad, which they are not, nor are men with similar ears." Moreover, he allows them to be "lively and facetious, and when they fall in a passion their ears redden and move. And sometimes their temples redden, which is a sign of modesty." But if the ears are flat and small, with the curled parts smoothed out, "they are cruel like the rustic dog."

The ear depicted is square at the bottom

with a small lobe, and the orifice is much too small. The pointed top is formed by a long straight Division (1) slanting upwards at half a right angle. Division (2) is very small, and is like a little cap to the point. Division (3) is absent. Division (4) is thin and perpendicular, bulged below, where it runs into Division (5). This Division (5) takes a sudden angle in order to go nearly horizontally to meet the lobe.

### 10. *Women's ears*

Auribus esse solet mulier mediocris honesta
Prudens et quæ sit moribus egregiis.

(An honest prudent woman is wont to be moderate in the size of her ear, and such may be of the best of good manners.)

Ghirardelli only admits one form of ear for women, and that is of moderate size. He seizes the opportunity to describe his ideal woman. "For a woman to be prudent, modest, and wise, she should not only have ears of a moderate size, but also her figure, and with a square forehead of just measure, her cheeks somewhat plump, with a natural colour, a pure speech, voice between grave

# DELLA PORTA AND OTHERS 81

and sharp, eyes large and brilliant, long and nervous hands, swift of foot but light and graceful. In fact, she is full of every grace and majesty. She is conformable in her actions, well-advised and diligent in the ordering of her house, never does a ridiculous thing, smiles little although she understands fun, and speaks less, very modest, loving solitude, despising vain ornaments and dresses simply, her temperament is cold and dry and melancholy."

For once the artist has drawn a handsome face and a pretty little ear. It is neatly shaped, oblong, with a rather large lobe and an orifice nearly the right size. The *helix* is delicately folded over. Of the Five Divisions, Division (1) is large, Division (2) is moderate, Division (3) runs into Division (4), which is long and slender and joins Division (5), which has a slight bulge near the lobe. It is so nearly natural an ear, that it was probably drawn from the ideal woman herself.

On the whole, Ghirardelli seems to have taken the most pains to observe for himself. But it is needless to say he had no scientific basis for his observations, and that his results partook somewhat of the miraculous. Occasionally he describes with sufficient accuracy for us to be able to place the ear more or less under a modern category, but the results are much more prosaic than his. We have had more careful analysers of character since those times, whilst the introduction of photography has given us exact shapes to catalogue. The long lists of shapes of ears and their attendant qualities remind us of the courtly fashions of old, when kings and queens went about properly crowned and having attendants holding up their trains. Sometimes imposters wore those crowns, but the attendants still held up the trains. So the ancient lists of ears expect to be accepted as perfection and the qualities as inevitably true. Unfortunately no logic was present as master of the ceremonies, and the result is an odd jumble of ideas, forming the vaguest possible sketches of character, often of impossible characters themselves.

Dominico de Rubeis of Venice (1639) was one of the most careful compilers of this kind. He drew up a set of tables in Latin, giving thirty-six kinds of ears, large and small and medium-sized, long, short, square or round. If we take these thirty-six kinds thrice over for the three sizes, and four times again for the four varieties, we obtain four hundred and thirty-two various kinds of ears. About ten qualities go to each ear on an average, so that after all we have less than five thousand specimens to choose from. It did not occur to this or to any otomorphologist that every ear differs from every other, hence that each is a specimen in itself. Even two centuries ago the general notion of relative numbers was very inaccurate. Statistics relating to anthropology were governed by astrology or guess-work, and whether a nought or two were added at the end of any statement in figures depended more upon what the compiler thought probable than upon investigations. Rubeis doubtless believed his list was exhaustive, and he had no idea of the number of people in the world who would have to go without a place in his tables. He seems also to be well satisfied as to the characters shown by them.

We subjoin one or two specimens of his nomenclature, because he has evidently written in good faith and with unusual succinctness.

amplæ $\begin{cases} \text{(Aures significant)} \\ \text{audacem, vanum,} \\ \text{impudentem, sed longævum.} \end{cases}$

That is to say, large and wide ears indicate the owner to be "courageous, rash, impudent, *but* long-lived." This little word "but" brings up a picture of the period of Charles I. of England, and of the state of the city of Venice two hundred and sixty years ago, when to be "courageous, rash, and impudent," commonly led to sudden broils and early deaths. Rubeis himself seems surprised that the man with a large wide ear should ever escape with his life and reach old age. He regards with more certainty and complacency the chances of the man whose ears are only moderately large, and who has therefore a good memory which retains things well, and who is also wise, honest, strong, *and* long-lived—

magnæ modice $\begin{cases} \text{bonæ retentionis et memoriæ,} \\ \text{sapientem, probum, robustum} \\ \text{atque longævum.} \end{cases}$

Rubeis has a weakness for little ears that are thin—

parvæ subtiles { probum, fidum, justum, pacem amantem, secretum, memorem mediocrite, timidum, tenacem, bonæ indolis, (s)atque docilem.

Yet there is a mixture of qualities we cannot admire in this " honest man, faithful, just, a lover of peace, very secret, with but a moderate memory, timid, tenacious, of good character and docile " It seems to be almost as dangerous to have small ears as wide ears —memory and courage being of so poor a sort.

These extracts are enough to show the kind of way in which otomorphology was studied at that period. It was never as favourite a subject as the face alone, because it was beyond the domain of astrology, although it did not escape the interference of the prognostications belonging to the " temperaments." And when wigs became universally worn, the ears went out of fashion, so to say, and nothing more was heard about them for more than a century.

# CHAPTER VII

### LAVATER ON THE EARS

THE quaint descriptions of ears already given, culled from ancient sources, are chiefly useful to show the futility of non-scientific work on the subject. With the eighteenth century a new method was adopted by the celebrated physiognomist, Johann Caspar Lavater (1741-1801). He was a Swiss pastor of a Protestant church in Zurich, Switzerland. In his youth he went to Berlin, with Fuseli the artist. He wrote "Swiss Songs," and numerous other works in prose and verse, and his sermons were admired and printed. Lavater was a friend and correspondent of Goethe, who much admired him, and submitted to his physiognomical criticisms. Goethe wrote: "Lavater's spirit was altogether imposing. . . . One amongst the most excellent with whom I have ever attained to so intimate a relation." Lavater was shot at by a soldier when Zurich was taken by the French in 1799, and after prolonged suffering of more than a year he died of the wound in 1801.

Lavater was of benevolent temper, and he devoted himself to physiognomy with the spirit of an artist caught in the new magic of science. Hitherto science (Lat. *scientia*) had meant simply *knowledge*, but now it had taken the meaning of a particular way of obtaining a special kind of knowledge, by careful observation and recorded research. The book by which he is chiefly known is his folio work in four volumes, profusely illustrated, called *Physiognomical Fragments, towards Promoting the Knowledge of Men and Human Love.*

Lavater could not help beginning by reading the ancient so-called authorities we have named, but he soon began to observe for himself, and to make or obtain portraits selected for the features he required to illustrate his theories. There is a peculiarity in his method which is apparent on a careful examination of the hundreds of faces in these volumes, and of the judgments he gives on them with energy and with inexcusable vagueness; it is that he judges every one by a rigid Swiss standard whatever may be their nationality. Little alterations of feature due to race alone are seized upon as proving a whole new theory of character, and this is in spite of his endeavour to give the faces of various nations as typical instances of racial form. The Swiss nation contains both German and French populations, welded together by their unique position among high mountains, and by their original form of government. They present strong national characteristics, shown in their faces with greater freedom—though not with greater play of feature—than is usual in other countries. Therefore, whenever Lavater tried to judge other nations' physiognomies, he read them from his Swiss standpoint, and this led him to give emphasis to the wrong parts. It is this unacknowledged cause which has often made him to be discredited, because, with all his painstaking details, he is so often in error in their application. Besides, he is full of exclamatory admiration and disdain, without explaining what has caused these emotions further than to say, "Observe the delicacy of the profile," *undeveloped*, as we should rather call it; or, "The ear is remarkable" (*merkwürdig*), without a reason adduced, when it is of an ordinary kind, ill-drawn, and not even matching the slope of the nose. The difficulty of winnowing out any grain from this heap of chaff is very great. Occasionally some feature is sufficiently well drawn to be recognisable, and by chance the character attributed to it may be distinct enough to be tested. Allowance must be made for the Swiss deflection of his judgment—like that of the needle from the North Pole—and then the information can be applied for the reading of modern photographs. After some practice the eye gets trained sufficiently to observe living features, and the patient student is surprised to find that Lavater is right about once in five times, a percentage far in advance of any physiognomical system ever invented up to that time.

It was owing to the difficulty of making use of Lavater's observations that his system never got developed as a whole. He seems to have had an instinct for judging character, such as novelists should possess, which was all-sufficient to himself, but which eluded employment in accurate description. Being a pastor, he had opportunities of knowing the characters of his flock that were probably useful to him with regard to individuals, but the range was not wide enough for him. When he took to the study of portraits, he did not make enough allowance for inaccuracies in portraiture. Lavater complained gently at times that a profile would seem to be " ill-drawn," and that it was " unnatural " and " impossible," yet when he chances to point out the parts he considers to be " impossible," further investigation from real life is apt to prove their existence, and that the shape had evidently been selected by the limner as a beautiful and uncommon form.

The extreme difficulty in presenting the shape of the ear in drawing, painting, or sculpture is emphasised in the portraits given. Even when Lavater draws attention to a drawing of an ear as being worth inspection the shape and position are incredible to a modern otomorphologist, and the lobe seldom tallies with the chin. He does not appear to have got much help from ancient writers—which is not surprising—and his own best attention was usually bestowed upon the rest of the face and figure. In the first edition of his well-printed folios, published by subscription in Leipzig, 1775,[1] he gives one short page about ears, with five full-sized and shaded drawings, with the following remarks :—

Up to now we have said nothing about ears in a general way. But it has already often been remarked that there is a special physiognomy of the ears. I am, however, still so far behindhand that I have extremely little to say about the matter. This time, therefore, I shall speak only about a few ears. Here is a page with four.

Not one is that of an extraordinary man. Not one of a man who works with courage and zeal, right and left, high and low, near and far, with ease compelling notice!

The first ear is that of a common, weak person. The second, I should certainly say, is more open and decided. The third is of a person who has very great capacity for learning and for teaching, as if meant for a schoolmaster or precentor, not a mere heartless smatterer in knowledge. The lobe is much more delicate at the bottom than those of the rest, especially the fourth. As to the fourth, I almost think it must belong to an

---

[1] *Physiognomische Fragmente, zur Beförderung der Menschenkenntniss und Menschenliebe,* von Johann Caspar Lavater. Leipzig und Winterthur, 1775 (Bod. Lib. Oxford).

extremely weak head. The broad flat ear, without a rim at the top, might otherwise show an excellent genius—and is especially observable in many musical ears—but here the whole has such a pervading flatness and coarseness and spreading out (I speak from the copper-plate lying before me) that I very much doubt if ever an innate genius could have such an ear. The ear below this notice is too decided to belong to a coarse person, and too round to belong to an extremely cultivated man.[1]

We may now briefly describe the ears.

*Ear* 1.—Division (1) is short and thick. Division (2) is long, arched, and thin. Division (3) is absent. Division (4) is well-shaped, bulged, and tapered into Division (5). The lobe being thick and wide denotes a large square chin. The orifice is moderate in size.

*Ear* 2.—Division (1) is too shapeless to define. Division (2) is arched and shapeless. Division (3) is very long. Division (4) is very short. Division (5) is long and thick. The lobe is very wide and thick, indicating a heavy wide jaw. The orifice is of medium size.

*Ear* 3.—Division (1) is long and thick.

[1] Ears 3, 4, and 5 are again given in Lavater's later "Additions," under the heading of "Addition C, Ears 1, 2, and 3." I have given a more detailed analysis of them, together with a photograph (taken by permission from the copy of Lavater's book in the British Museum) in the following chapter under the later heading.

Division (2) is drawn with a double bulge of an arbitrary and unnatural kind. Division (3) is bulged and distinctly good in form. Division (4) is bulged, and it runs into Division (5), which is long. The lobe is rather pointed, which indicates a pointed chin. The orifice is wide, and about the right size.

*Ear* 4.—Division (1) is enormous, both thick and bulged outside. It runs into Division (2), which ends by the *anti-helix*—what we should call the "flap" of the ear,—spreading out to the edge and leaving no further space for any *helix*. There is no rim to the rest of the ear. This ear *is* wide, and the lobe *is* short and wide, indicating a wide jaw with horizontal jaw-bone and a moderate chin. The orifice is wide, but too small.

*Ear* 5.—Division (1) is thick and bulged. Division (2) is short. Division (3) goes into Division (4), which runs into Division (5). This coalition of divisions is bulged at Division (3). The lobe is full, indicating a thick chin. The orifice is too small and shapeless.

In later editions of his book Lavater says more about ears. We quote from the English edition (vol. iii. Part II. ch. ix. p. 411) :—

I frankly acknowledge that this subject is yet rather new to me, and that I pretend not to pronounce with full assurance concerning it. In the meantime I am fully convinced that the ear, as well as, perhaps more than, the other parts of the human body has its determinate signification, that it admits not of the least disguise, that it has its suitableness to and a particular analogy with the individual to whom it belongs. All physiognomical study ought to be founded on exact drawings, on comparisons and approximations frequently repeated. With regard to the ear, I would advise you to pay attention (1) to the totality of its form and size; (2) to its interior and exterior contours, to its cavities, and the hollow of it; (3) to its position. You must observe whether it adhere close to the head or be detached from it. Examine this part in a brave man and in a coward, in a philosopher and a changeling-born, and you will soon perceive distinctive differences referable to each character. In the vignette below I do not perceive one single form which I could suspect of stupidity. I even believe them all to be above mediocrity, and that which is in the centre most probably supposes a sage and luminous mind.

This vignette we reproduce. It holds twenty-one ears, prettily though not very accurately drawn, and all at first sight appear to be much the same. Lavater does not tell us, for evidently he does not know himself, where to look for these indications of a "sage and luminous mind," or how to prove that none of the number could be suspected of stupidity. Probably it is the presence of a distinct *helix* in every ear that averts his suspicions, and the squareness of several of the ears would attract his favour, as the presence of a *helix* and any squareness of shape

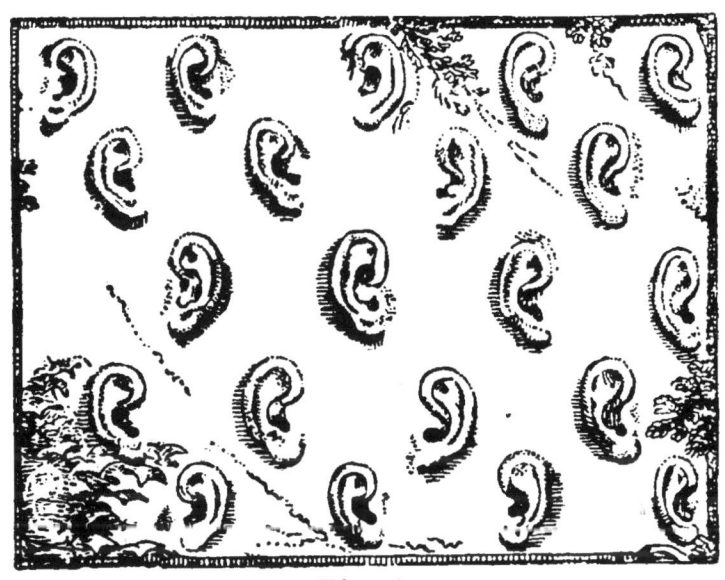

Fig 4.

were both much approved of by the ancient physiognomists, some of whom Lavater refers to by name. It would be of use to know whether these ears are portraits, or are drawn from casual recollection or from mere fancy as to what ear the wise man should possess. They are certainly not common ears, as they have very odd collocations of the Five Divisions of the *helix*. These are indeed so arbitrary in arrangement that it is scarcely possible to believe these ears ever existed in the shape given. For however strangely nature may deal out the Five Divisions, there is a certain method in her "assortment" wholly wanting to the imaginary ear.

(Fig. 4 is photographed by permission from Lavater's work at the British Museum.)

A quaint little show-boy hides one ear with his arm, and a special part of the *helix* of another ear is hidden by his leg. These two ears have therefore been omitted in the following description, and are not represented.

We can now describe these ears, taking them in rows from left to right, beginning at the top row.

*Ear* 1.—Division (1) is short and thick, much curved, and joining Division (2). Division (3) is absent. Division (4) is long and well shaped, nicked slightly where Division (5) begins, which is very long and thick, running down to a large pendent lobe. The orifice is rather narrow.

*Ear* 2.—Division (1) is long and thick, making a point where it joins Division (2), which slopes down into Division (3). The rim is thin all round, except a bulge in the upper part of Division (4), which is very short, and where also the *pinna* widens outwards and then curves downwards to the big lobe. The orifice is drawn too small.

*Ear* 3.—Division (1) is short. Division (2) is thick and long, flat at the top and forming square joinings to Division (1) and where Division (3) should be. But Division (3) is absent, and a long straight Division (4) begins at once from the joining to Division (2), and it has a squared nick where it joins Division (5), which itself runs down into the long and pointed lobe. The orifice is tiny.

*Ear* 4.—Division (1) is shapeless and curved. Division (2) is short and curved. Division (3) is absent, and in its place the *helix* is nicked and square. Division (4) is long, bulged, slightly curved, and well shaped. Division (5) is short. The lobe is large. The orifice again is tiny.

*Ear* 5.—Division (1) is scarcely shown, and runs into Division (2), which is short and slightly flattened. Division (3) is very long, and the *pinna* slopes outwards here. Division

(3) runs into Division (4), which is short and bulged and runs sloping back into Division (5), which is thick and long. The lobe is very long and pendent. The orifice is shapeless.

*Ear* 6.—Division (1) is absent. Division (2) is short, thin, and slightly curved. Division (3) is very thick and long. Division (4) is very short, thick, and bulged. Division (5) is short and thick. The lobe is long and pointed. The orifice is tiny.

*Ear* 7.—Division (1) is short and thick and with a square inner corner where it runs into Division (2), which is short and thick. Division (3) is very tiny. Division (4) is straight and short, and runs into Division (5), which is thick and long. The lobe is heavy and very wide. The orifice is medium-sized.

*Ear* 8.—Division (1) is short and bulged, nicked in a point, where it runs into Division (2), which slopes rapidly down, absorbing Division (3) and running into a thick, short, bulged Division (4). Division (5) is short. The lobe is heavy. The orifice is tiny. The *pinna* is very narrow and long.

*Ear* 9.—Division (1) is shapeless. Division (2) is bulged and pushed backwards towards Division (1), making a point at the end towards Division (3). The *pinna* then slopes gradually outwards to the middle of the *helix*. Division (3) runs into Division (4), which is thick and long and runs into Division (5). The lobe is very large and wide and deep. The orifice is very narrow and upright.

*Ear* 10.—Division (1) is short and with a square corner, where it joins Division (2), which has also a square corner where Division (3) should be, but is absent. Division (4) is very long, curved, and thick, tapered into Division (5), which is thus prevented from being further shaped. The lobe is very large and pendent. The orifice is small.

*Ear* 11.—Division (1) is short and squared where it joins Division (2), which is straight and thick and long, and is squared where Division (3) should be, but is absent. Division (4) is very large, long, bulged, pushed too high, and sloping outwards to the middle of the ear. Division (5) is too high, and is short. The lobe is extremely large, thick, and pendent.

*Ear* 12.—Division (1) is shapeless. Division (2) is small. Division (3) is short and thick, running into a large, bulged, thick Division (4), that goes into a thick Division (5). Where Division (5) ends, it is bulged outside. The lobe is long and pendent  The orifice is rather small.

*Ear* 13.—Division (1) is long, thin, and upright, with a square corner where it joins Division (2), which is very short and pointed upwards where it joins Division (3). The *pinna* is long and narrow. Division (3) is short and straight and joins Division (4), which is short, bulged, and pushed too high. It joins Division (5), which is thick and short and bulged outside at the bottom. The lobe is long and pendent. The orifice is medium-sized.

*Ear* 14.—Division (1) is thick and shapeless. It pushes up into a point where it joins Division (2), which slopes down and has a square corner where Division (3) should be, but is absent. Division (4) is tapered above, and is very long, bulged, and thick below, where it pushes out like an elbow where it joins a long thick Division (5), sloping into a long, heavy, pointed pendent lobe. The orifice is medium-sized.

*Ear* 15.—Division (1) is short, straight, upright, and pushes up Division (2) into a point where it joins it. Division (2) is long, straight, and thick, it slopes down to where Division (3) should be, but is absent. There is a square nick where Division (4) begins, which is long, thick, straight, and squared outwards where it joins a short thick Division (5). The lobe is heavy and pendent. The orifice is small. This ear is like the preceding one, except that Division (4) is shorter and straighter.

*Ear* 16.—Division (1) is almost absent. Division (2) is short and curved. Division (3) is absent. Division (4) is high up, long, thick, bulged, and tapered below. Division (5) is short and sloping into a large pendent lobe. The orifice is tiny.

*Ear* 17.—Division (1) is short and shapeless. Division (2) is short and thick, pushed backwards down into Division (1) by its pointed top. Division (3) is short, thick, and straight, and helps to finish this point, and then slopes into a long, thick, straight Division (4). This runs into a short, thick Division (5) with a square bulge outside. The lobe is wide and heavy. The orifice is very small.

*Ear* 18.—Division (1) is short, thick, upright, with a square corner where it joins Division (2) which slopes downwards and is squared at Division (3). This runs into a long, bulged Division (4) which tapers into Division (5), thus preventing it being further formed. The lobe is short and wide, but thick. The orifice is also too small.

*Ear* 19.—Division (1) is short and thick, it pushes up into a point where it joins Division (2), which slopes down and tapers where Division (3) should be, but is absent. Division (4) is tapered, thick, bulged, long, and with a square corner where it joins a short thick Division (5). The lobe is extremely wide, heavy, and pendent. The orifice is square.

*Ear* 20.—Division (1) is short, thin, squared where it joins Division (2), which is short, thick, and bulged, and squared where Division (3) should be, but is absent. Division (4) is long, thick, goes straight down, but is bulged and runs into Division (5), with a slight nick showing where it begins; it is very short. The lobe is long and pointed. The orifice is small.

*Ear* 21.—Division (1) is short and thick, and squared where it joins Division (2), which is short and runs into Division (3). This goes into Division (4), which is long and slightly bulged and tapered, and where it joins Division (5) there is a bulge outwards. The lobe is wide and rounded. The orifice has not been drawn.

Out of these twenty-one ears, Division (4) is long in fifteen cases, and where it is short it has the sign of activity, *viz.* being thick or bulged.

# CHAPTER VIII

### LAVATER'S OUTLINES OF EARS

LATER on, Lavater adds a series of larger ears in outline, rather better drawn, and occasionally almost life-like. His comments on them are unfortunately of the same confused nature, and can only be accounted for by his having read the old writers—such as della Porta—carefully, and differed from them by instinct, without method or reason. As we have seen, they strongly insist on the necessity of a distinctly curved *helix*, and of the advantage of square bits about the ear. But they did not sort the shapes of the *helix* at all definitely, and when they drew ears the result was not satisfactory. Lavater was very conscientious and painstaking, but he could not draw well enough nor get sufficiently accurate draughtsmen for the work. Nothing short of photography or nature-printing can give the shape or proportions of the ear in a manner fit for scientific use, or even for physiognomical study.

In his chapters of "Additions" on the ear, Lavater says:—

Having made so little progress in the study of the ear, it will be difficult for me to give a positive and satisfactory commentary on the additions made to this chapter. The comparison of extremes will, in time, furnish me with inductions more certain; I believe, however, I run no risk in affirming that among the designs of the annexed plate you cannot find a single one which characterises imbecility.

*Ear* 1 appears to me the most delicate, the feeblest.

*Ear* 2 is more ingenious, more attentive, and more reflecting.

*Ear* 3 is superior to 1, with respect to activity and energy. I think I perceive in it a productive genius, rich in talents, and, particularly, endowed with that of eloquence.

I adopt nearly the same definition for No. 4, but with some modifications, the reason of which I look for in the upper part. On the other hand, the serpentine contour which binds the hollow may probably be the sign of good-nature.

*Ear* 5 is much weaker, and more contracted than 2, 3, 4.

*Ear* 6 is still smoother, and less undulated. I except, however, the point which is under the hollow, and which, in spite of mediocrity of faculties, seems to indicate a particular talent, I know not what.

According to my text, *Ear* 7 announces a man modest, humble, and gentle, perhaps timid and apprehensive.

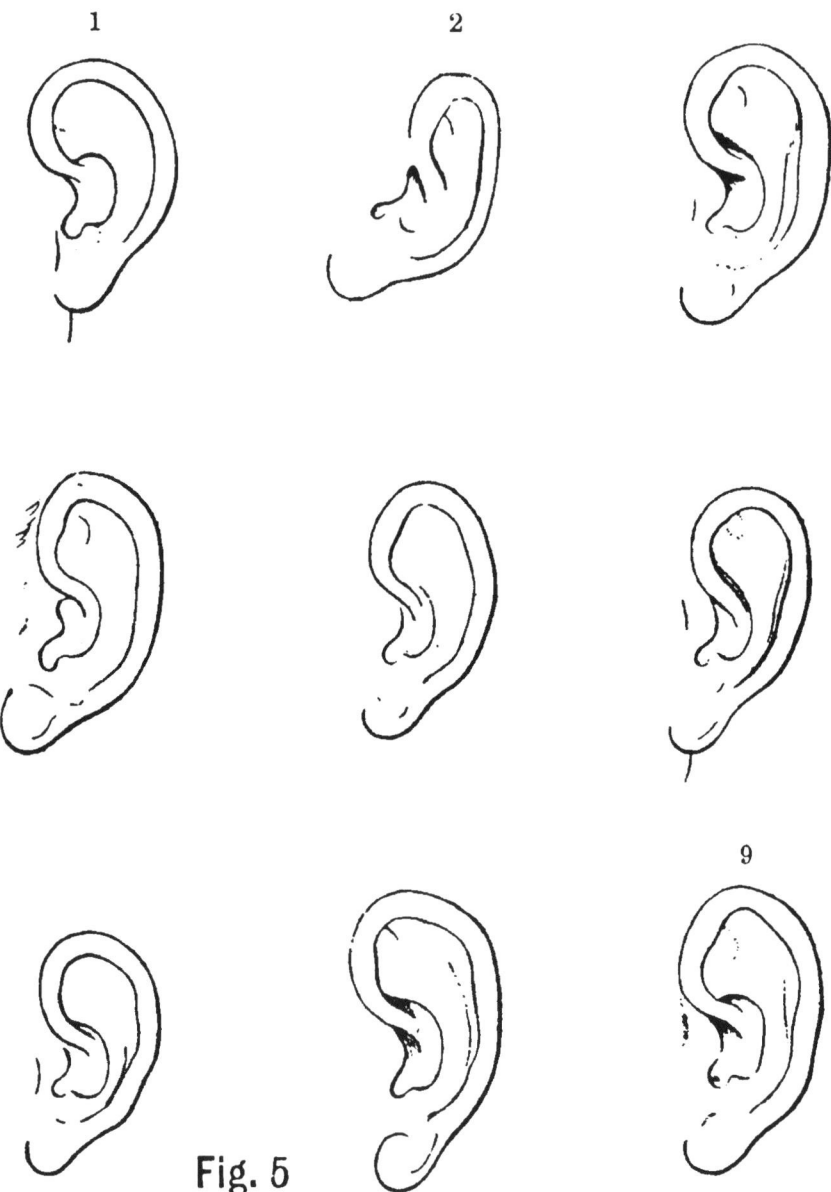

Fig. 5

*Ear* 8, and still less 9, cannot possibly belong to ordinary minds.

It would be interesting to bring together a hundred different and well-known heads, and to abstract from them the proper and specific character of their ears. In those under review the tip is disengaged, which may always be considered as a good omen of the intellectual faculties.

(These nine ears are photographed by permission from Lavater's work in the British Museum.)

By describing these nine ears after the method I have used in the previous chapter, we can bring them into range for purposes of classification, identification, and physiognomy. It will be seen at once how very incorrectly they are drawn.

*Ear* 1.—Division (1) is thick and much curved, it thins a little as it runs into Division (2). The *helix* then remains the same thickness throughout Division (2), which slopes down into Division (3) and Division (4), until it reaches Division (5), where it is bulged a little outside. The lobe is long and hanging. The orifice is very narrow and upright. The curve outwards of Division (1) is so marked that it brings the whole *helix* much too forward on to the cheek.

*Ear* 2.—This is a specimen shaped like

those of Ghirardelli's "dog-ear'd men." The lower half of the *pinna* is broad and nearly square, the lobe is pushed up as if lying on its back, it would be large and pendent otherwise. The rest of the *pinna* is very narrow and upright. Division (1) is very large and bulged, it goes nearly straight up and partly absorbs Division (2), which itself slopes suddenly down into a long narrow Division (4), upright and a little bulged. Division (3) is absent. Division (4) runs into Division (5), which makes a kind of elbow of itself and is also a little thicker than all the other divisions, except Division (1). The orifice is meant to be of the wide shape, but it is drawn only half the right size.

*Ear* 3.—Division (1) is large, curved, and much bulged. Division (2) is thick and curved but short, and runs into Division (3). Division (4) is long and bulged and well shaped, joining Division (5) and pushing it downwards. Division (5) is thick and short. The lobe is almost round, and hanging. The orifice is upright and too small.

*Ear* 4.—Division (1) is long, upright, bulged, and tapered below. It joins with

Division (2) in a sort of thick peak, for Division (2) begins at once to slope down. Division (2) is thick and medium in length, and it runs into Division (3) and Division (4), pushing the latter downwards. Division (4) is upright and very slightly bulged, it is of good length and is tapered below. Division (5) is short and thick. The lobe is long and hanging, but pushed along a little. The *pinna* is altogether narrow and upright. The orifice is much too small.

*Ear* 5.—Division (1) is long and bulged, tapered below  It runs into Division (2), which is short and curved, and runs into Division (3) and Division (4). The *helix*, except at Division (1), is the same width all round, but there are two very slight nicks, which show where Division (4) begins and ends. It is pushed up high by a very long Division (5), which curves out where it joins Division (4), and again curves out below as it slopes to the round and hanging lobe. The orifice is much too small and is upright. The whole *pinna* is less sloped than usual.

*Ear* 6.—Division (1) is long and bulged above, it runs into Division (2), thus making the top of the *pinna* rather less curved and almost flat. Division (2) is very short, and it runs into Division (3), which is large and bulged. Division (4) is long and tapered above and below, it is bulged, and it is pushed much too low by Division (3). There is but little room left for Division (5), but it is bulged. The lobe is long, oval, and hanging. The orifice, as usual, is too small, it is upright. The *pinna* is of a rather narrow form.

*Ear* 7.—Division (1) is long, rather straight and narrow. It runs into Division (2), which is slightly thicker and is curved, and runs into a short Division (3). Division (4) is short and bulged, and tapered above and below, it is pushed too high, to make room for a long thick Division (5), which is bulged below. The lobe is small and hanging. The orifice is square and large, and drawn quite out of place.

*Ear* 8.—Division (1) is very long, tapered, bulged, rather curved; it runs into Division (2), which is thick and sloped into Division (3) and Division (4). A slight bulge in Division (4) shows it is well placed; it is upright and a good length, and slightly tapered as it reaches Division (5), which is short and bulged. The lobe is very large, long, and hanging. The orifice is very narrow and upright. The *pinna* is rendered too long in proportion to its width, on account of the large lobe.

*Ear* 9.—Division (1) forms an elbow with a thick bulge, tapered at each end, and it runs into what should be Division (2), but this is hardly more than a bend joining Division (3). This latter is tapered and also very long, thick, and bulged, not only taking up what should be most of the space for Division (2), but also running down so far that it pushes Division (4) too low, which is itself rather straight, short, and bulged. Division (5) is short and thick, and bulged outside. The lobe is wide and pendent. The orifice is not much too small.

Lavater obtained the outlines of twelve more ears, and wrote of them as follows :—

>Addition B.
>Twelve Ears.
>Each of these forms varies in its length and cavities, in its exterior contours, and the hollow in the middle. Each is adapted to only such and such a head; each bears the impress of an individual character.

*Ear* 1 is likewise the first in rank for gentleness, simplicity, modesty, and candour.

*Ear* 2 is more undulated, more susceptible of cultivation.

*Ear* 3 is still more delicate, more sprightly, and more attentive than the preceding two.

I confidently maintain that 4 cannot be the ear of an ordinary man, but is perhaps a little more harsh than 3.

*Ear* 5 is probably the most original and most lively of the twelve.

*Ear* 6.—More phlegmatic than 3, 4, 5, with less sensibility than the last, but of much greater capacity than 1.

*Ear* 7.—Replete with wit and ingenuity.

*Ear* 8.—The rounding of the upper contour is very singular. I know not what to make of it; only I doubt whether this ear has the merit of the preceding.

I suspect 9 of a little timidity; in other respects, I ascribe to it justice and activity.

*Ear* 10 appears to me insignificant, inconsiderate, volatile, and insipid; its facility is mere shuffling.

*Ear* 11.—Circumspection destitute of every species of courage.

*Ear* 12 scarcely admits of violent passion. I discover in it modesty and gentleness, founded on dignity of sentiment.

Following the same arrangement as before, we now give a photograph of these twelve outlines of ears, and proceed to their analysis.

(These ears were photographed by per-

## LAVATER'S OUTLINES 113

mission from the work by Lavater in the British Museum.)

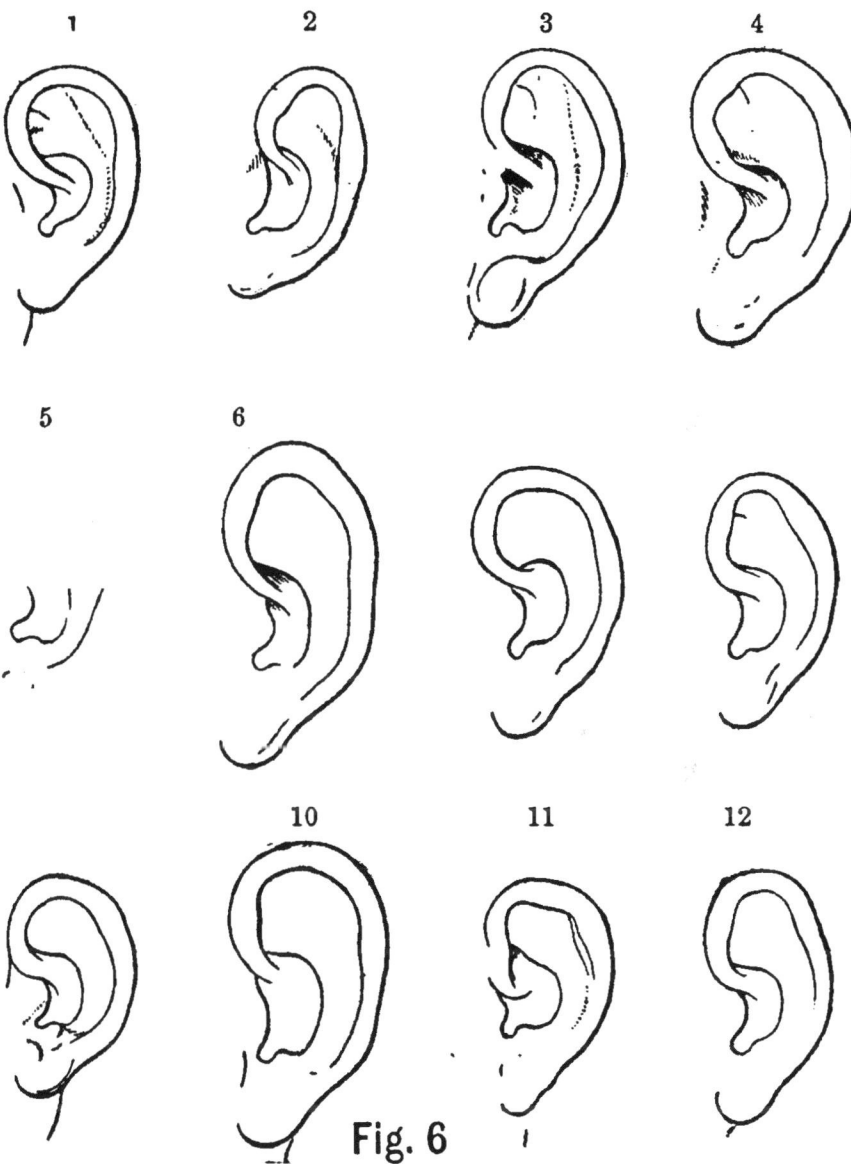

Fig. 6

TWELVE EARS (FROM LAVATER)

*Ear* 1.—Division (1) is very thick and

curved, tapered below, but a little thinner above, where it runs into a long flat Division (2) of the same breadth. This latter runs into Division (3), which is short and thick, and slightly bulged, and it pushes Division (4) downwards. Division (4) is rather short and bulged. Division (5) is scarcely formed. The lobe is long and hanging, but is not thick. The orifice is drawn too small.

*Ear* 2. — Division (1) is large, bulged, and sloped up to Division (2), which is short and very curved, and runs into Division (3). The *pinna* is thus made rather narrow at the top. Division (4) is long, and bulged slightly inside and a good deal outside. It is pushed rather low, and tapers slightly into Division (5), which has a slight bulge where it slopes off to the lobe, and another bulge just before touching the lobe. Although the lobe is really nearly as long as usual, it is very slightly marked in consequence of these bulges in Division (5), so near to it. The orifice is very upright and very narrow, though rather square at the bottom.

*Ear* 3.—Division (1) is large, bulged, and nearly upright. It runs into Division (2),

## LAVATER'S OUTLINES

which is narrow, rather short, and not much curved. The *pinna* here slopes suddenly, as Division (2) runs into Division (3) and a rather bulged Division (4). By this means Division (4) is pushed too high in order to absorb Division (3) above, and to leave room below for a large Division (5), which is bulged and tapered, beginning above the middle of the ear. There is a very long, thick, hanging lobe. This is what would be commonly called a "usual" form of ear, and it is the normal size—that is, the length is double the width, but the orifice is drawn much too small.

*Ear* 4.—Division (1) begins very low down and bulges, whilst curving very much outwards. Division (2) is thick and bulged, pushed backwards on to the top of Division (1), and then it makes a sort of flat bend towards Division (3), into which it slopes and is absorbed. Division (3) has only a slight indent to show where it runs into Division (4), which is short, thick, and bulged inside and outside, and is rather high. It runs into Division (5) with an indent to show where Division (5) begins; this is thick and bulged outside above and below, and also slightly bulged inside. The lobe is thick and hanging, but it does not show its full length, as Division (5) runs so low down towards it. The orifice is too small.

*Ear* 5.—Division (1) is short. Division (2) is pushed down on it at one end, and then slopes suddenly upwards into a kind of peak. Division (2) then completes this peak by sloping suddenly downwards into Division (3), which is long and thick, bulged outside, running into Division (4). The latter is bulged both inside and outside. The *helix* makes a sort of elbow where it joins Division (5), which begins high up at the middle of the *pinna*, and it is long and thick, and bulged below outside. The lobe is small, pendent, and rather pointed. The orifice is much too small, rather square at the bottom, but very narrow.

*Ear* 6.—Division (1) is tapered, very long, and rather upright, and is thick and bulged both inside and outside. It runs into Division (2), which is thick and moderate in size, and goes at once into a thick, sloping, bulged Division (3). This runs into a long Division (4), which is nearly upright and of medium thickness. It runs into Division (5), which slopes off to the lobe. This lobe is long, thick, and hanging, and is pushed somewhat slanting. The orifice is too small.

*Ear* 7.—Division (1) is long and not thick, and rather squared as it starts, and then it runs into Division (2), which is long and rather flat, and it makes a sort of square corner at Division (3). This latter is short and bulged outside and tapered into Division (4). In this ear Division (4) is noticeable for being very well formed, exactly in place, gently bulged in the middle, and tapered above and below; it runs into a short, bulged Division (5). The lobe is long and hanging, but not very thick. The orifice is too small as usual, and the bottom slopes up. It is a rather small ear.

*Ear* 8.—Division (1) is thick as it starts, and it goes nearly upright, and is long, bulged, and then tapers as it makes a sort of square corner in joining Division (2). This latter stops in the middle, and Division (3) begins and runs into Division (4); this makes the *pinna* very narrow at the top. The lobe is thick and hanging, and of medium length The orifice is too small and slopes upwards.

*Ear* 9.—Division (1) is thick below, then it narrows as it goes up, and widens again by absorbing part of Division (2). The top is not much curved, and it runs quickly into Division (3), which is pushed too high and is bulged outside. It runs into Division (4), which is also bulged outside and also pushed too high, and which runs into Division (5), bulged below. The lobe is very round, though hanging. The orifice is very small, as usual, and too narrow. The *pinna* is much less in length than twice its width.

*Ear* 10.—Division (1) is bulged both sides, and is long and thick. It goes nearly straight up into a large Division (2), which is rather thick and flat, and runs into Division (3). There is a slight nick where the latter joins Division (4), which is upright and gently bulged, long, and tapered into a short bulged Division (5). The lobe is long, thick, and hanging. The orifice is not drawn so small as usual, but it is somewhat too narrow.

*Ear* 11.—Division (1) makes an elbow at its start, and then another as it joins Division

(2). This latter is long and thick, flat-topped, and bulged inside. The *helix* is well nicked where Division (3) should be, but is absent. Division (4) begins, therefore, too high; and it is thick, bulged, and tapered below into a long, thick Division (5), that begins above the middle of the *pinna*. By this means the *pinna* is square in the upper part, but slanting below. The orifice, though bigger than usual, is still rather small, and is square at the bottom. The lobe is very pointed, with an outside bulge near the top.

*Ear* 12.—Division (1) is very long, bulged, and nearly upright. It tapers into Division (2), which is narrow and not long, and runs into Division (3). The latter runs into Division (4), which is slightly bulged. It slopes outwards, and tapers vaguely into an indefinite kind of Division (5). The lobe is hanging, and is rather rounded, though long. The orifice is too small and slants upwards.

A final "Addition C" is supplied by Lavater, who seems to have here only kept the three last out of the five ears given by him in the early edition of his book, quoted in the previous chapter :—

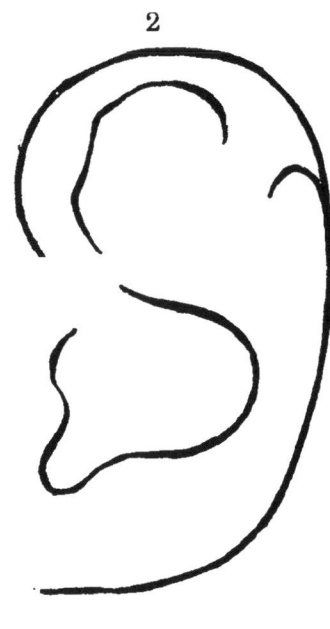
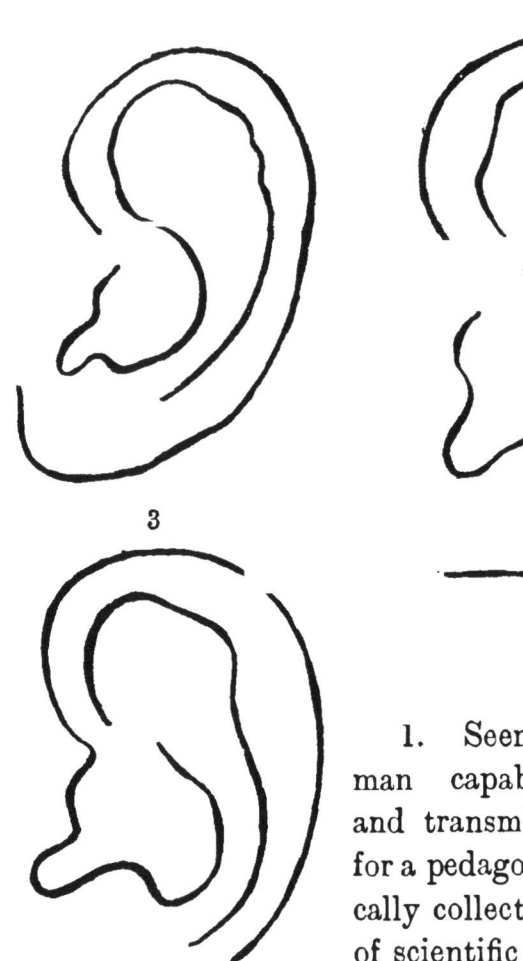

Fig. 7

1. Seems made for a man capable of acquiring and transmitting knowledge; for a pedagogue, who mechanically collects a great number of scientific articles.

2. Cannot be referred to any but a head excessively weak. This form broad and smooth; this want of rounding in the contours may in truth subsist with superior faculties, and particularly be found with musical ears; but when the whole is so flat, so coarse, so tense, it certainly excludes genius.

3. Has too much precision to ascribe it to a blockhead; but, on the other hand, it is too round and too massy to furnish the indication of an extraordinary man.

(These three ears from "Addition C" have been photographed, by permission, from the work by Lavater in the British Museum.)

These ears are worth a more detailed analysis than that given hitherto, on account of their peculiar construction.

*Ear* 1.—Division (1) is long and thick, bulged both inside and outside, and tapered into a small Division (2), which runs at once into Division (3), indicated as the latter by a bulge outside. Division (3) runs into a long Division (4), which has some jags in it. We must notice the difficulty of classifying this ear; for if it had not been for the bulge outside Division (3), the top jag of Division (4) might have been mistaken at first sight for a pushed-down Division (3). Division (4) tapers slightly into a long, thick Division (5), that is bulged below both inside and outside. The lobe is small, pointed, and hanging. The orifice is wide and about the right size.

*Ear* 2.—Division (1) is very large, curved, thick, bulged outside, and it goes up so far that it pushes Division (2) out of place, making a tapering where the top of Division

(2) should be. Division (2) is very short, and, in fact, joins the *anti-helix*, so that there is no more *helix* all round the rest of the *pinna*. The ear is wide both above and below. The lobe is wide, straight, and narrow. The orifice is almost square and is too small.

*Ear* 3.—Division (1) is thick and bulged both inside and outside; it joins Division (2), which is very thick and bulged below, and sloped towards Division (3). The latter is only indicated by the thick joining with Division (4), which is bulged and tapered below into Division (5). This last is of medium thickness, and seems to be a roll upon the ear rather than a rim, and it slopes, and slightly tapers towards the lobe. It is this kind of rounded Division (5) which makes the ear noteworthy. The lobe is rather pointed, short, hanging, but wide and thick above. The orifice is odd in shape, the bottom being curled as if deformed from a blow; it might be termed shapeless, and it is also too small.

# CHAPTER IX

### THE PHYSIOGNOMY OF THE EAR

THE characteristics shown by the ear form a sort of physiognomy of their own. It is a peculiar kind of physiognomy, for it does not mention the usual qualities seen in the face, but it gives a mere list of capabilities without saying whether they are used or not. And though there are very few in this list, the combinations of the qualities are so numerous and varied that it would require a mathematical calculation to compute them. Yet they can be all ranged, as they occur, under the rules given in Chapter II These characteristics appear to be congenital, indicating certain qualities waiting to be developed. As the ear appears to a casual observer to be beautifully and perfectly formed from infancy, it has been supposed never to alter. We have already noticed that M. Bertillon assumes this as probable in his book on *Identification*.[1] But, in fact, the ear grows and adapts itself to the growing face.

It is very curious indeed to watch the ears of children altering slightly in shape as they develop their bodily and mental features. The size of a baby's ear in comparison to its mite of a nose, and rose-petal of a face, will hint as to the future development of that nose, and the lobe will suggest what shape the tiny chin will take. An infant can be identified by its ears with the greatest ease for about two years. After that they gradually alter, more often in the actual shape and size of the *pinna* than the way in which the rim is folded over. The ears are sometimes flat without any *helix*, but this is not of frequent occurrence. The rim has been known to curl over of itself at the age of two weeks, and to tighten the fold later on. Such a case was brought to the notice of the present writer.

Although the ear indicates qualities waiting

---

[1] The exact words are: "Quelques-unes des variations de forme que présente cet organe paraissent subsister sans modification depuis la naissance jusqu'à la mort" (*Identification anthropométrique*. Instructions signalétiques, par Alphonse Bertillon, 1893, § 47, p. 66).

to be developed, we cannot tell which of them will be unfolded in the future. Still, if the lobe is large in an infant's ear, we may expect the chin to grow large, and, therefore, that the full-grown face will have a short nose. But if the whole ear which has a large lobe is already large and long, we shall not only expect to see a large chin, but also a long nose. The face in growing larger will fit itself to the ear, and the ear will grow longer if the face happen to become elongated; and the *helix*, in that case, will alter in its curves. We may, therefore, make a rough guess as to which parent the child will eventually take after in appearance. But yet the actual folds of the *helix* will show its own selection of qualities, and will possibly resemble neither parent's ears, but those of some ancestor, direct or collateral. Whilst the ear is thus altering, if we observe which relative's ears the child may be taking after at the time the stage of development can be understood, and a clue to the present form of character will be found,— extremely useful to those who have to guard and guide it.

We give an example of the alteration in

shape of a child's ears from the age of two years and a half to that of five and a half. The following is the analysis :—

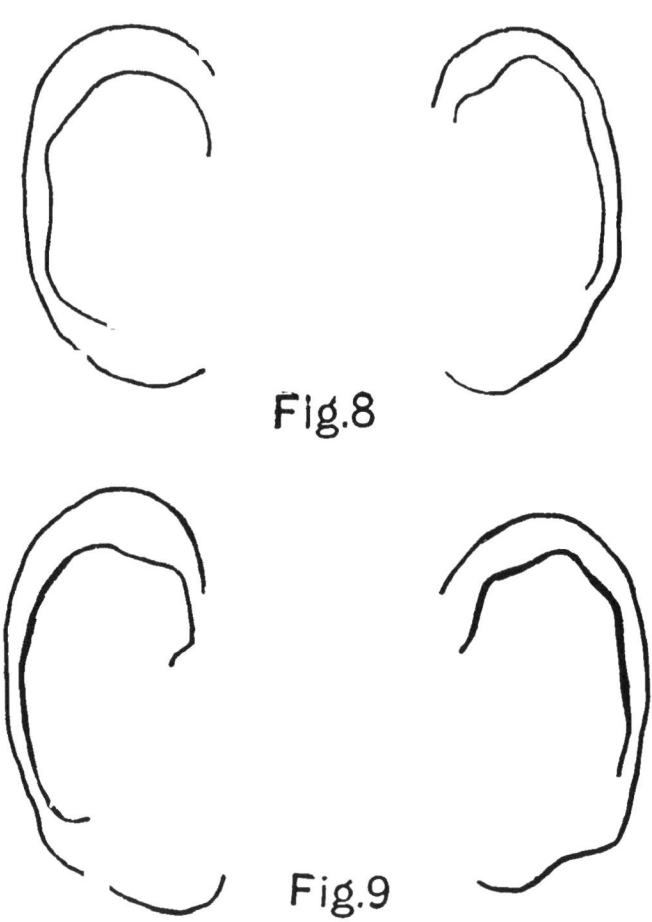

Fig. 8

Fig. 9

At two and a half years old the *right ear* was as follows (Fig. 8) :—Division (1) is long and thick, and runs into Division (2). This latter is thick and runs into a long, thick Division (3), which tapers where Division (4)

# PHYSIOGNOMY OF THE EAR

begins. This Division (4) is well shaped, tapered above and below, and is of good length. It has been pushed rather low by the long Division (3). Therefore, Division (5) has to begin lower down than its right place, and it is short and thick, and runs into the wide lobe, which is gently curved.

At two and a half years old the *left ear* was of the following shape (Fig. 8):—Division (1) is short. Division (2) is thick and bulged, and pushed backwards on to Division (1). Division (3) is only indicated by a slight squareness, as the *helix* slopes downwards and then outwards with a long, slender, slightly bulged Division (4). The latter tapers slightly as it runs into a long and slender Division (5), which is bulged below, and then makes a slight elbow as it joins the wide lobe, itself of a moderate size.

Comparing the above with the following (Fig. 9), we find the alteration as below:—

At five and a half years old the *right ear* has altered a good deal. Division (1) has separated from Division (2) and is now short and thick and nicked where Division (2) begins. The latter is large and well curved, and pushed a little backwards into the place of Division (1), and at the other end it runs into Division (3) as before, but Division (3) has grown larger with the lengthening of the *pinna*, and instead of tapering to Division (4) it absorbs Division (4), pulling it upwards. Division (4) is thus a good deal shorter, besides being altered in shape and position, though it still tapers down into Division (5), which has a short thick elbow close to the lobe. The lobe remains the same size as before.

At five and a half years old the *left ear* has thickened both in Division (1) and Division (2), whilst Division (3) has enlarged and tapers to Division (4). This Division (4), owing to the increased length of the *pinna*, has had room to lengthen downwards till it is now unusually long. It tapers into Division (5), which is flat and makes a square bulge outside, a sort of elbow. Although the *pinna* is enlarged, as is the case with the other ear, the lobe is about the same size as before.

We find that the tops of the ears are developed in the same direction, and the lower portions are also developed, rather than altered. It is the side of each ear which has chiefly altered in these three years of childhood. That this is not wholly due to the mere enlargement of the *pinna*, but to its *systematic* enlargement in the direction originally intended by nature, may be partly gathered from the singular fact that the alteration in the *right* ear has made it grow like his father's left ear, whilst the *left* ear has become different from those of both parents.

At about fourteen years of age the ear appears to be full-grown, although the character is usually in a very juvenile condition. Yet the roots of the adult qualities have already struck deep, and it becomes necessary to choose a future career for the boy or girl. Nevertheless, however well we may be able to read their ears, it will be still quite as difficult to make a choice. The average ear is much the most common, as the name indeed implies, and with average qualities a youth can be trained to almost anything. Perhaps the chief point of use in the reading of the average ear, is that it saves the possessor from being expected to possess talents above the average. Now the average ear is possessed by astonishingly promising young folk. For it contains tendencies towards capabilities of every kind, together with the freshness of untried powers. Besides a great many genuine capabilities have to go towards making up a sound and useful character. The point to which we wish to draw attention, is that the ear will show which is the quality with the strongest staying power, so that it may be relied on to help the other powers to keep the whole up to the mark, without any exceptional talent. If the boy or girl has any desire for a special pursuit, this can be taken into account, without our being obliged to believe that this desire in itself indicates a talent. The ear that is wide and square at the lower part generally belongs to a more impressionable character than the narrow ear, and the owners in youth frequently desire a career entirely unsuited to their powers, simply from having a friend or relation whose achievements in that direction have dazzled them. Yet if they want to follow their father's profession, and if he happens to be able to assist them

# PHYSIOGNOMY OF THE EAR 131

to do so, then their impressionability comes in useful in making them eager to learn what they may not have any talent for, their one talent being, in fact, an adaptability to favourable circumstances. The narrow ear seldom adapts itself to anything.

There is no fortune-telling to be found in the ear, as we have already observed. Even the ancient magicians—often the most learned doctors of their time—could not get the stars and planets to look after this portion of the profile-face, and though these learned doctors of all the science of the period harped upon temperaments, they could not explain away this very obstinate member of the human frame. As we have seen, they thought men's ears resembled those of animals, and that, therefore, such men had the same characters. If we should try to tell animals' characters by their ears, we should have to invent a system for each kind, or we might invert the ancient process and say the animals had human qualities according to how their ears resembled human ears.

The following classification of capabilities by the Five Divisions of the *helix* are illustrated in each case by examples, taken from nature-prints :—

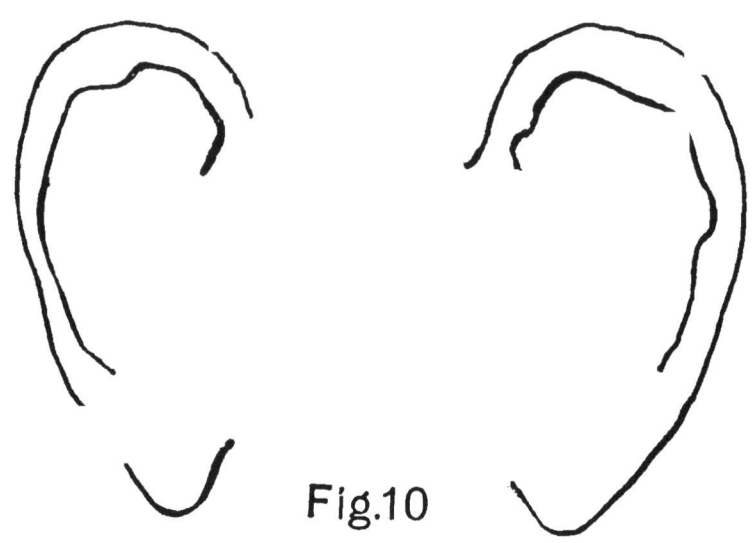
Fig.10

*Division* (1).—This refers to personal pride

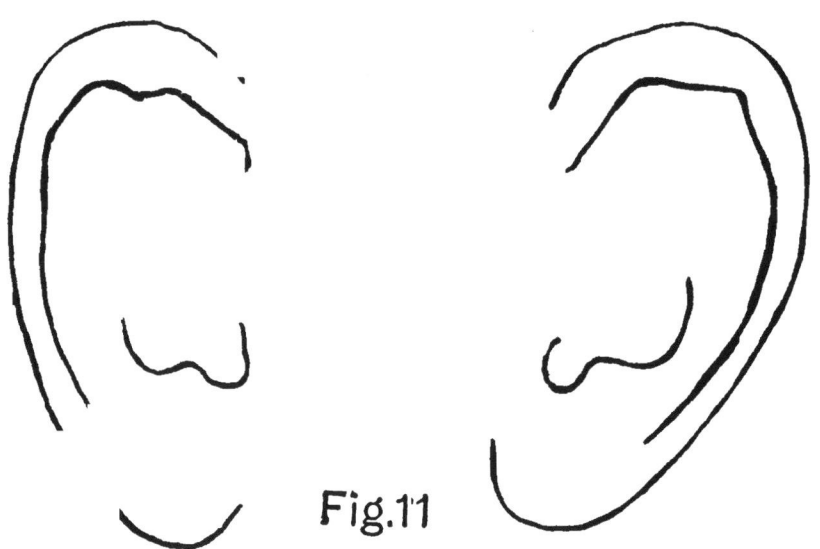
Fig.11

of the kind called "proper pride." The owner of this ear is fully aware of pos-

# PHYSIOGNOMY OF THE EAR 133

sessing this quality as a ruling passion (Fig. 10).

*Division* (2).—This goes with a tendency to unreasoned convictions, such as whims or prejudices, not necessarily good or bad. In this example the whim has taken the form of a special pursuit (Fig. 11).

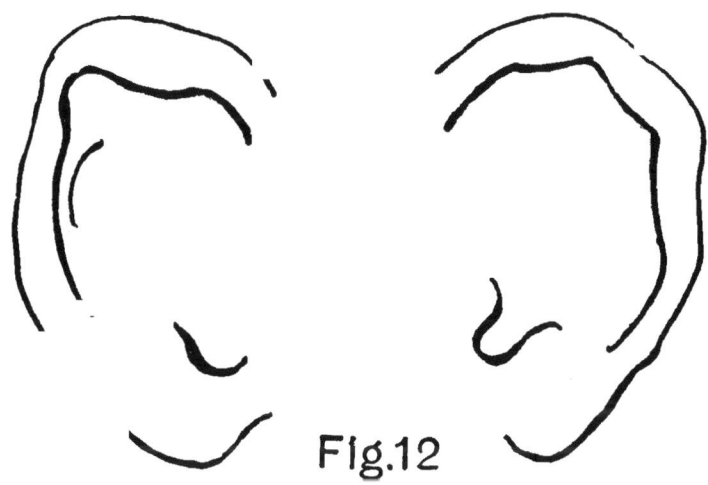

Fig.12

*Division* (3).—This indicates a liking for out-of-the-way subjects, which are distant in time or place, or for anything outside the usual run of interests  It is found with antiquarians, and with those who undertake original research, etc. This is an unusually fine example, where Division (3) is twice as large as its normal size in one ear, dominating it entirely, as the other divisions are of medium size (Fig. 12).

*Division* (4).—This belongs to a power of continued attention of the mind, but it does not indicate mind alone. It is often pushed upwards, as if drawn up by the tendency to a deficient Division (3), or sometimes as if to leave room for a large Division (5). On the

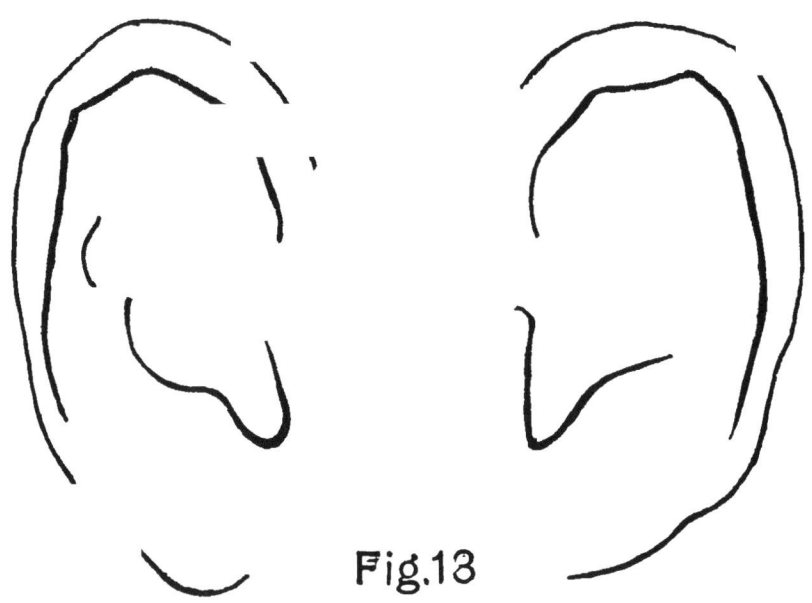
Fig.13

other hand, it is sometimes pushed downwards, absorbing part of Division (5), and thus leaving room for a well-developed upper part of the *helix* Its normal place (see Fig. 1) is from above the middle of the side of the *pinna* ending just below the centre of the ear (Fig. 13).

*Division* (5).—This gives a persistent

# PHYSIOGNOMY OF THE EAR 135

attention to material circumstances. How this quality is applied will be seen in examples later on (Fig. 14).

This short list must appear to consist of mere vague capabilities. Such as they are, they will be found to be distributed differently in each ear, and here the mystery of the

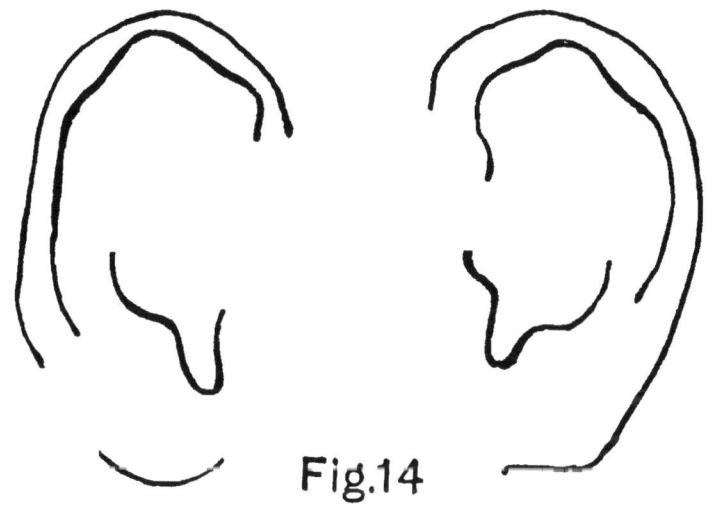

Fig. 14

working of these capabilities begins to be seen.

In explaining the method in which these Five Divisions can be usefully employed in the attempt to catalogue certain qualities, no assertion is made as to whether these qualities are good or bad. All that is claimed is that the probable manner in which the native qualities will be used is indicated in the ear.

Each ear announces the way in which one-half of an undertaking will be carried out. From the observation of many thousands of ears I am inclined to consider that whatever is placed in the right ear will show the method followed in the first half of an undertaking—that is, in right-handed people, but the reverse in those that are left-handed. This is only advanced as being possibly true, because I have never yet found an exception to the rule. But this is not so important as the particular fact that where the ears differ, the undertakings will vary in their course. Very seldom are a pair of ears found which are a perfect match, and hence a good deal of time and anxiety would be spared if the owners of the usual unmatched ears could guess betimes which way they were likely to vary in their undertakings. This makes at once a wide opening for the unrolling of character, and the best way to make use of one's qualities.

For instance, if Division (1) is absent in both ears, the personal pride does not lead the character, but if it appears in one ear, half of every undertaking is tinged with this quality, while if it is present in both ears, the whole character is dominated by it.

Again, with Division (2) the whims are abandoned halfway through if only marked in one ear, or they are not taken up until the undertaking is half completed. The word *whim* is not here used in a bad sense. For the capacity of taking up a subject with a ready-made conviction, such as a whim undoubtedly is, may be often helpful when the undertaking happens to be worth carrying out in the teeth of difficulties. A prejudice in favour of a thing is another form of whim, being an unreasoned but ardent conviction, which will often succeed where the lack of it would cause failure. Prejudices against a thing, however, need not be deduced from the presence of a large Division (2), unless there is no safeguard against this form of whim in the rest of the ear. This part of the ear varies a good deal. Sometimes it is pushed up into a point by Division (1) being too big, or else pushed backwards into a short thick Division (1), giving a prick-cared shape in either case. Sometimes it is as flat as a ruler at the top, absorbing Division (1) and

Division (3) on each side of it. The normal shape is a gentle arch. The position merely modifies the way the whims are set about.

With Division (3) the tendency to delight in things of the past is chiefly found amongst the upper educated classes. Archæology and antiquarianism in some or all of their branches are the most common pursuits. It gives also a bent towards out-of-the-way and ancient languages, or philology in connection with the ancient parts of modern languages. It goes also with the search after *curios,* or any thing or any subject apart from the beaten track of the owner's life. It is absent in more than 50 per cent, and when present it is seldom found in more than one ear. So strong is its influence that when it is present in both ears it shapes the career of the owner in spite of himself. The particular quality it indicates in any given ear may be surmised from comparing the shape and size of the other divisions of the *helix.*

The special power of fixing the attention on any subject, shown in Division (4), works in very unforeseen ways. It does not make people clever, yet clever people are seldom

## PHYSIOGNOMY OF THE EAR 139

without some of it, and really stupid people—if such exist—will have some more or less futile form of it, overbalanced by one of the other divisions which drags off the attention to its own quality. Yet extremely stupid people may possess Division (4), if they persist in their stupidity with sufficient diligence. Very clever people can be entirely without it, if persistent attention is not required for their form of talent. It can be seen in abundance at all meetings of learned societies: amongst the industrious of those who attend Extension lectures; in the *habitués* of the British Museum reading-room; and at afternoon "teas" amongst noted housewives. Where it is only present in one ear the owner is either precocious, and a subsequent failure in life, or else accounted stupid in youth and unexpectedly clever in middle age. What to make of such a capability lies apparently in the power of the possessor. Skilled workmen have it, and members of Parliament who are hard-working representatives of our country. There is something akin to pleasure in the spectacle of a man with a feeble Division (4) making a speech where most of his audience listen with ears well dowered with long, bulged, and tapered forms of Division (4). The dawning astonishment on their faces at being expected to accept such inadequate representations as are being forced upon them, followed by involuntary and unanimous laughter, succeeded by well-timed repression of the speaker by the chairman or president, at last goaded into Olympian though restrained wrath, and the spluttering of the repressed speaker, like a cheap lamp turned down too suddenly—all testify to the value of the physiognomy of the ear in relation to Division (4). A very common form of Division (4) is when it is run up into Division (3), and also down into Division (5), without any break or nick, and the *helix* is sometimes very thick. The way in which their owners pursue their objects through thick and thin is amusing, because they do not stop to think a thing out properly for itself, but only regard it in connection with their own wants. They are limited, in spite of apparent intelligence, by this devotion to their own well-being. If this composite shape is slender and tapered top and bottom the owner requires early instruction to be well educated, but even then a lack of logic and of memory of anything outside his own ends will usually be found. However, if Division (1) is large, these ends will be in accordance with "proper pride."

The functions of Division (5) are numerous and varied. The owner likes well-being in every form, but not necessarily for himself alone. Hence comes the power of discovering what is lacking in well-being, together with the energy to supply it. Among other things Division (5) gives hospitality, and renders researches successful through attention to necessary material details. It also carries out schemes of benevolence, and does not disdain to take and give pleasure in fashion and dress. The power to deal habitually with authority with large masses of people goes with a well-shaped form of Division (5). Where the *helix* is shaped like a round roll instead of a folded-over rim, the owner instinctively aims at using or discovering laws of all sorts, whether abstract or concrete, that deal with large masses of people, or of ideas. When the *helix* is bulged outside and is straight inside the force of any division is weakened, and the converse strengthens it. A bulge on both sides gives overflowing energy.

We require but a slight knowledge of the laws of ethics to guess how the Five Divisions may clash and counteract one another, and to select the characteristics that can be deduced from the whole combination. Emotion and goodness *per se* are not to be found in the ear, nor intellect, nor wickedness. The only possible hint at any one of them is in the proportion of one division to another.

As all the specimens given in this book are selected from the ears of the upper educated classes, it is interesting to note how the capabilities marked by the Five Divisions work in them. Some of them belong to self-made members of society, risen through sheer force of talent and rectitude. Others belong to members of ancient or titled families. Many are celebrities, and others are of eminence in their respective careers.

To complete the physiognomy of the ear, we must now refer to the remaining points to be observed.

Wherever a little knot occurs, that division

# PHYSIOGNOMY OF THE EAR 143

possesses a power of specialising (Fig. 15). This ear has very little *helix*, yet it belongs to a talented woman who has two specialities apart from her usual avocations. They are marked by the arrows.

When the ear is wide at the base of the opening, it shows a power of appreciating

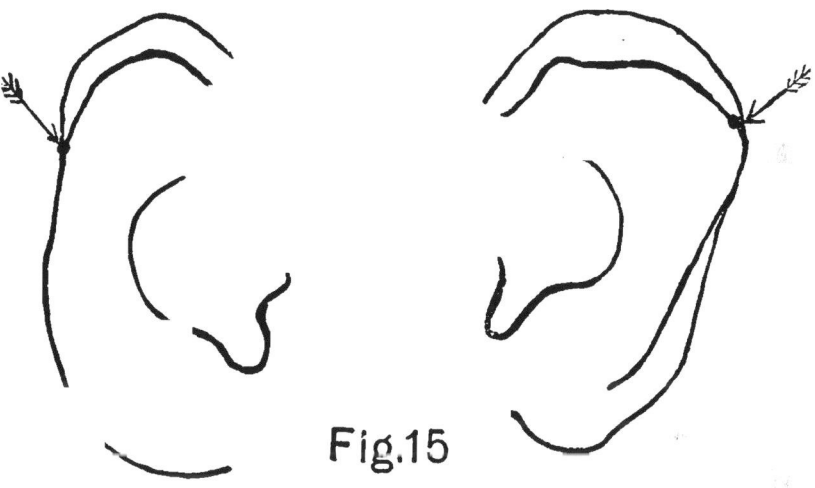

Fig.15

differences of sound. We give an example (Fig. 16) in the ears of Sir John Stainer, Mus. Doc., lately Professor of Music in the University of Oxford. The square shape is identical with the form of Mozart's ears in the best portraits of that great composer. It will be noticed that the left ear has a slightly squarer opening than the right ear, and it is very remarkable that, in comparing them with Mozart's, the same difference of shape is found, with the noticeable fact that the orifice of Sir John Stainer's left ear resembles that of Mozart's right ear, while the orifice of the right ear of the former repeats that of the left ear of the latter.

The square form is seldom absent in the

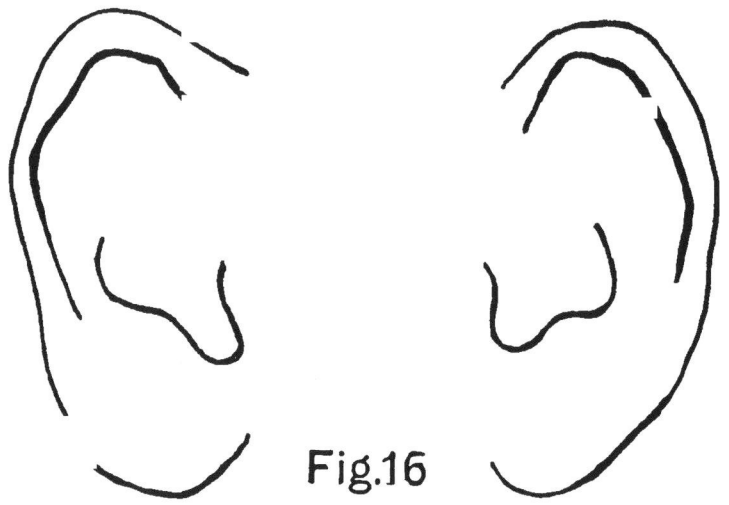
Fig. 16

orifices of the ears of musicians, especially composers, but though it brings delight in sounds it does not always belong to musicians. It is seen in large numbers in the best orchestras and at concerts where classical music is played, and amongst linguists it is not uncommon in a modified form.

When the orifice is much narrowed there is a tendency to deafness that should be guarded against. There is also a delight in sheer noise as being a thing cheerful in itself, probably because all the slighter cheerful sounds are unheard.

The ears occasionally stick out of the head, like handles of a jar. With such ears there will be less of the forehead seen, owing to the low-growing hair belonging to the type. The eye misses the face-colour, and the ears supply the deficiency, giving an amiable expression lacking to the side-face. If there is a sort of neck to the *pinna*, the projecting form is natural, otherwise we may be sure it has been caused by head-gear in childhood. The ear with a neck indicates a delicacy of physical health apparently, but it does not determine longevity or the reverse.

When the ear is pushed outwards at Division (3) so as to make the top of the ear as wide, or wider than, the middle of the ear, there will be found a natural capacity for expression in words, often, though not always, found in preachers, orators, or authors. Where it is lacking, the difficulty has to be overcome by practice if it is imperative for the person to speak in public or to write books, owing to his position or special knowledge. It is more common in men than in women, and this perhaps accounts for the former's zeal in oratory and leaders in newspapers. The patient talking for talking's sake, which is a feminine trait—intended for the transmission of languages to children from their earliest age—is not shown by any squareness of the ear, and must not be confused with this shape. In fact, the women who may chance to possess this shape of ear are likely to take to writing for the press or lecturing in public, as an unceasing flow of oratory is not appreciated in the domestic circle even by children.

# CHAPTER X

### FURTHER EXAMPLES OF THE EAR

THE variations of the shape of the ear and of the rim and of the orifice, we have shown can be classified both for identification and for

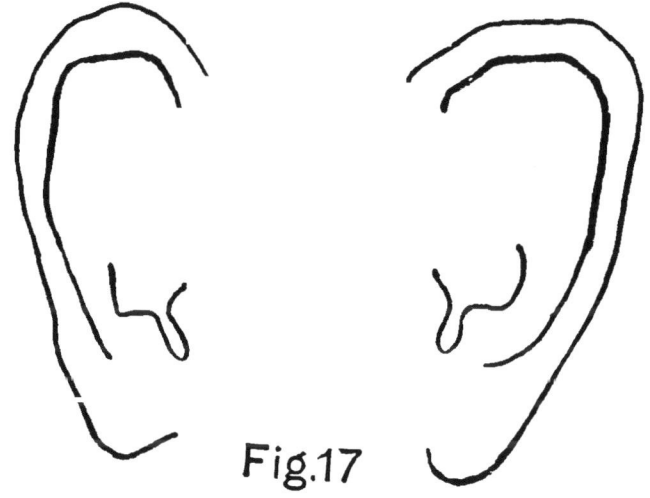

Fig. 17

physiognomy. In fact, the latter helps to fix the shape in the memory, making identification more quick and easy. We give some more examples of the way these rules can be applied, together with further modifications and their indications of capabilities.

When the top of the ear is straight (Fig. 17), there is a kind of independence of character that is often found amongst travellers. This level shape is in a very distinct form in the left ear of Fig 17. It belongs to Mrs.

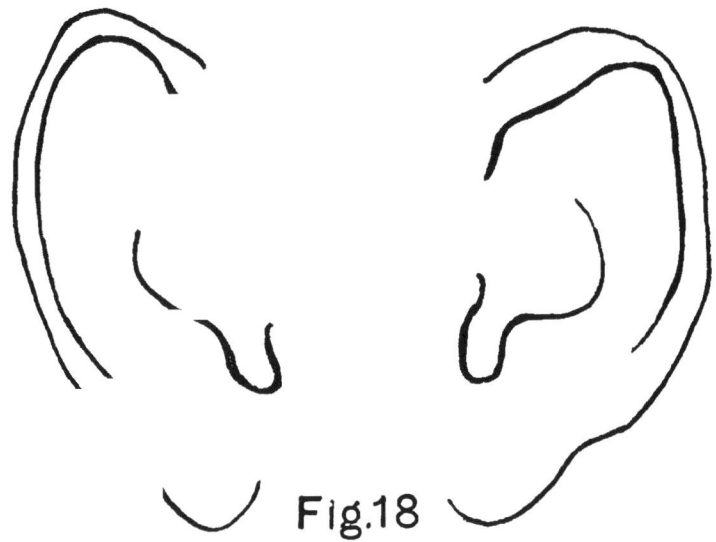

Fig. 18

Theodore Bent, the widow of the distinguished traveller in the Greek Archipelago, in Mashonaland, Arabia, and the island of Socotra. She accompanied her husband on all his journeys, which were often very arduous undertakings.

A gentle curve is the most usual form of the top of the *pinna* (Fig 18). The very high pointed form, caused by Division (2) being pushed out of place, which we noticed in the last chapter, seems to show the great impressionability which belongs to the artistic nature.  This is shown in Fig 18, where Division (2) is very pointed in the right ear. In the left ear, Division (2) is pulled back by a long thick Division (1) that goes straight

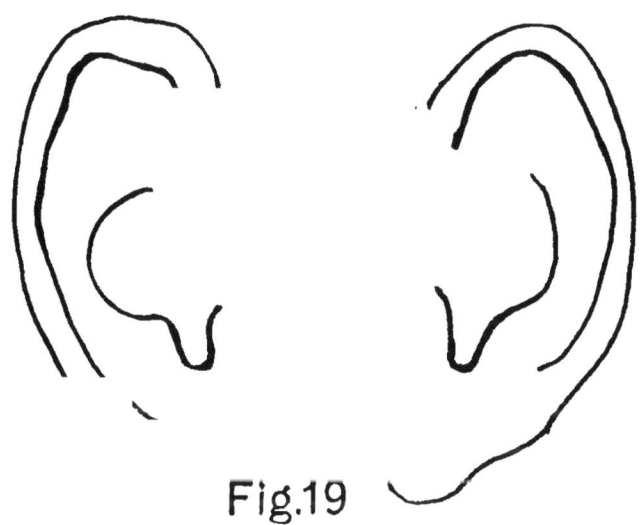

Fig.19

into it, with a nick at Division (3).  In both ears Division (4) is slender and joins a long Division (5), which is of good shape.  These are the ears of an artist, Miss **Annette Elias**, whose landscapes in the Royal Academy and other exhibitions show how impressionable she is to the beauty of English scenery.  The power of attention in Division (4) is directed to the resources of her art as shown in Division (5).

An interesting example of how these two forms of Division (2) act together is seen above (Fig 19). The right ear gives a traveller, the left ear an artist, the large Division (3) betokens a delight in out-of-the-way things, and the attention to details and masses of people is shown in the long Division (5). They belong to the well-known traveller, Mr. A. Henry Savage Landor, who describes himself as "artist and traveller," and who seeks the most out-of-the-way tribes and places to travel amongst and to sketch. Division (4) being slender, the attention is more devoted to action than to continued thought. The large Division (1) gives him self-reliance, which is a different quality from independence, and even more necessary under difficult and dangerous circumstances. It is noticeable that in the portraits of Mr. A. H. S. Landor before and after torture, the only feature that appears the same is the ear, and although it had become rather thinner it could have been at once identified (see *In the Forbidden Land*, vol. ii. p. 202). Owing to the head being more upright in the second portrait, the ear appears to be placed higher up. This illusion is dis-

# FURTHER EXAMPLES

pelled by placing a ruler even with the top of the ear and eyebrow in each case, when they are seen to be really in the same place.

(Fig. 20).—This ear is adduced to illustrate the way in which the Five Divisions are sometimes pushed about the *helix*, much as the features of the face are often out of proportion.

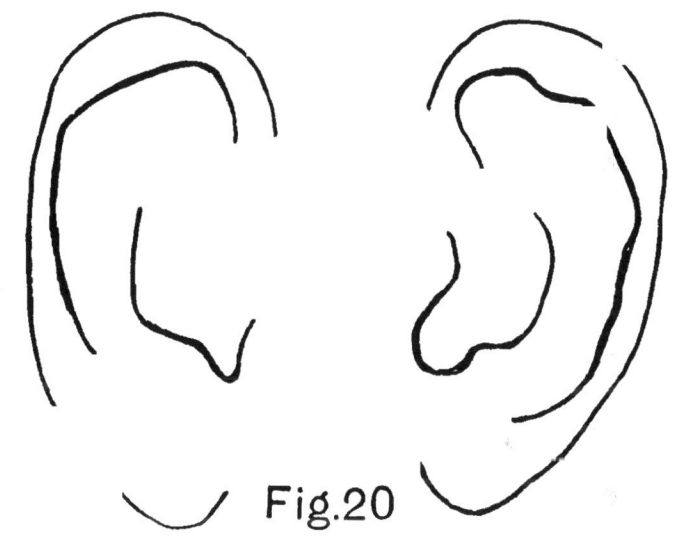
Fig.20

Here Division (2) is bulged and pushed down towards the place of Division (3) in the left ear, and then runs straight into Division (4), dragging it a little too high because Division (3) is absorbed between Division (2) and Division (4). Division (5) is tapered, beginning rather high, and it is of the form found when the owner has to superintend the welfare of others. In the right ear Division (2) absorbs Division (3) altogether, and Division (4) is not clearly shown except by its tapering top; it runs into Division (5), thickening the *helix* without a break, showing that the attention is used for emergencies belonging to Division (5).

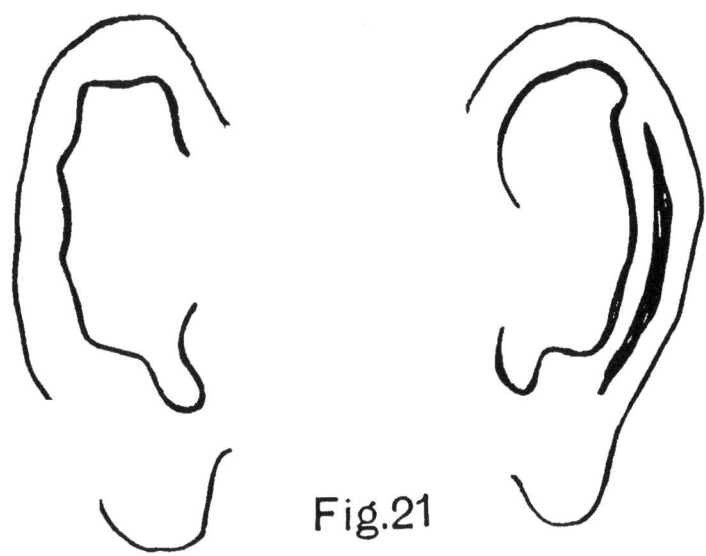

Fig. 21

(Fig. 21).—This is another peculiarity, viz. a double bend of the *helix*. It is deeply indented all down the side of the left ear, where the *helix* begins to bend over from where Division (3) should be, through Division (4) down to the end of Division (5). This in itself is so rare as to be of the greatest use for identification, whilst the right ear has

# FURTHER EXAMPLES 153

Division (3) so distinct, and Division (4) so short and bulged, and Division (5) so short and thick, that the two ears could not be mistaken for those of any one else.

There is nothing in Fig. 22 to show the nationality, which in this case happens to be

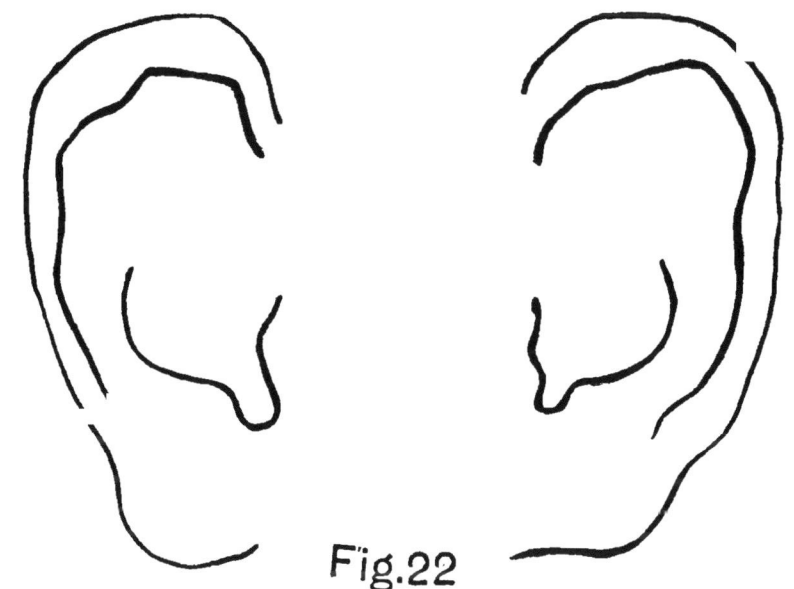

Fig. 22

American. We have not been able as yet to find that the ear has in itself a power of indicating nationality. For it shows individuality instead of race, whereas the other features reverse this method. In reading an ear we should class it as of a character resembling that of such or such a nation, which may be very different from the one to which the owner actually belongs. But how often we have met members of one nation whose characters strongly resembled those of another nation! Books of travel mention people who are found exceptional in their own nation. Such persons are attracted to the people of the other nation and frequently make their way to the country of their choice, showing

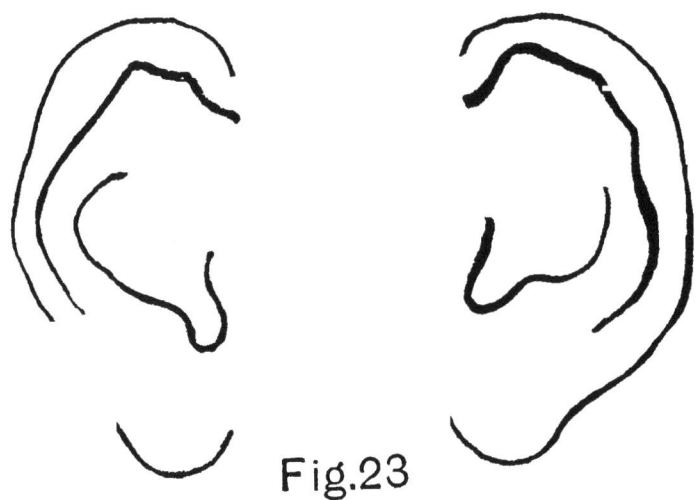
Fig.23

an unusual facility in learning and pronouncing the language.

(Fig. 23).—This an example of energy in exceptional circumstances. Division (3) absorbs Division (4) in the right ear and pulls it too high; and in the left ear Division (3) is as large as Division (4). Division (5) has a large place in each ear, and is well shaped.

# FURTHER EXAMPLES 155

Independence is shown in the top of the *pinna* being nearly flat. It is this, together with the size of Division (5), that gives the energy in looking after the well-being of many. The pair belongs to an Irish lady, a nurse at the Royal Naval Hospital at Haslar. Miss E. Keogh is one of the only two naval nurses who have received a medal. This was

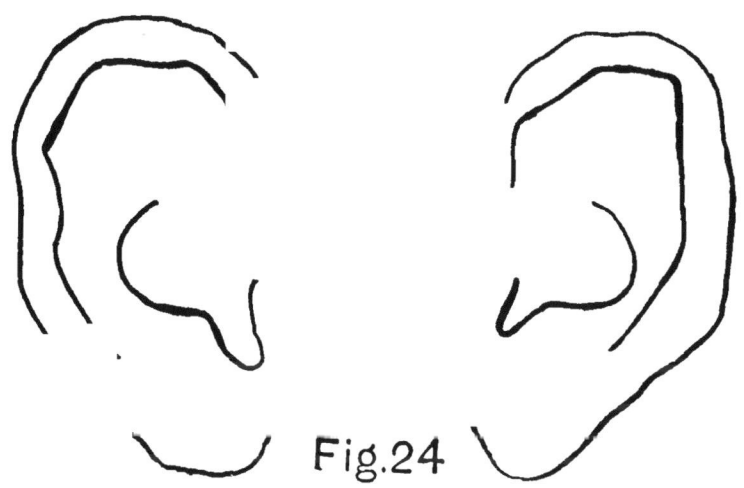

Fig. 24

for voluntary hospital service on board ship off the tropical coast of Benin during the expedition sent two or three years ago by the English against the natives of Western Africa. The ear is remarkable in itself, but might belong to any civilised nation.

(Fig. 24).—The peculiarity in the left ear is the long, thick, and straight Division (4), which is tapered above and runs into a short thick Division (5), betokening unusual powers of attention. By the addition of the large Division (3) in the other ear, the bulged Division (4), and the large Division (5), we may guess the owner has chosen some unusual fields of research in a strictly scientific domain, not unallied with powers of original discovery.

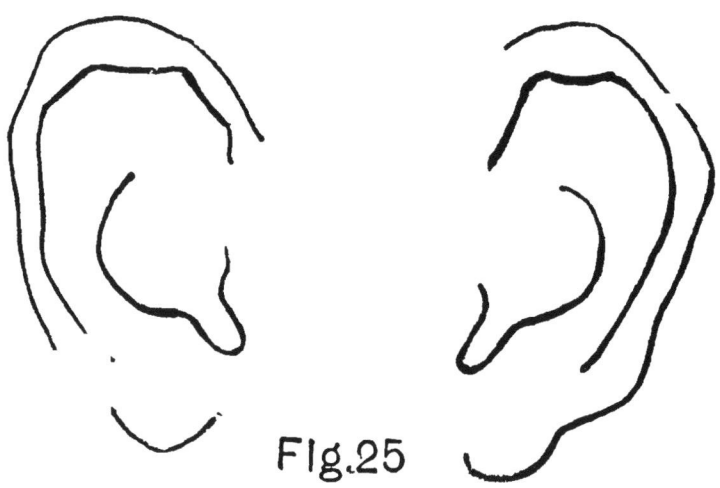

Fig. 25

Miss Florence Buchanan, of the University of Oxford, has already been known as a lecturer and discoverer in the science of zoology.

We give, in Fig. 25, a portrait of the ears of Miss E. E. Wardale, Ph.D. of Zürich, who, although English, has obtained this German degree, owing to her linguistic attainments. Division (3) is in both ears, showing interest in out-of-the-way subjects, and dominating the

## FURTHER EXAMPLES

use of the remaining four divisions. There is also the pear-like orifice that is often found with linguists, giving accuracy of hearing for verbal sounds. Division (4) being less developed and almost absorbed into the large Division (5) shows that the attainments are not used for further discoveries, but for others' well-being.

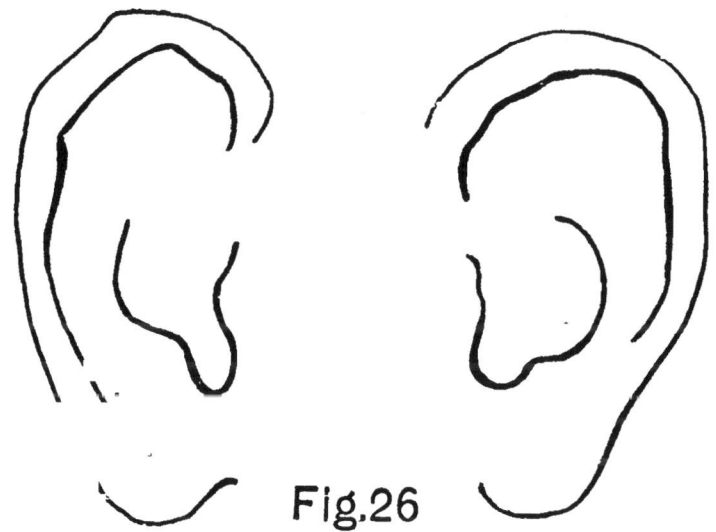

Fig. 26

(Fig. 26).—This is an example of how certain forms of ears seem made for special occupations. In the right ear Division (3) is so large that it takes up half the place of Division (2), which is pushed backwards upon the top of a short upright Division (1). This form of Division (3) shows scientific devotion

158    THE HUMAN EAR

to an out-of-the-way subject, which in connection with the pear-like opening of the orifice indicates linguistic interests. In the left ear Division (4) is long and tapered. Division (3) is short and of a compact shape, making the ear push outwards with a tendency to squareness, and the combination of Division (3) and

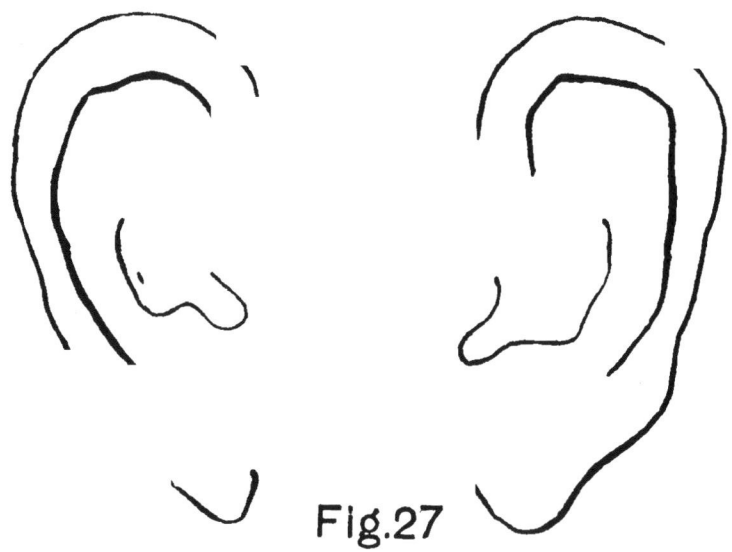

Fig. 27

Division (4), in these forms, shows an attention to the value of words. The lobe is large, giving firmness. This pair of ears, so admirably adapted to the selection and weighing of words according to their linguistic value, belongs to Dr. J. A. H. Murray, the editor of the great English Dictionary now being published by the Clarendon Press at Oxford.

## FURTHER EXAMPLES 159

We have hitherto dealt with pairs of ears where each ear differed more or less in a noticeable way from its fellow, for this is the most frequent case; and it is not easy to find a pair so much alike as those presented in Fig. 27. In both ears the rim is continuous, with only a slight hint of narrowing where Division

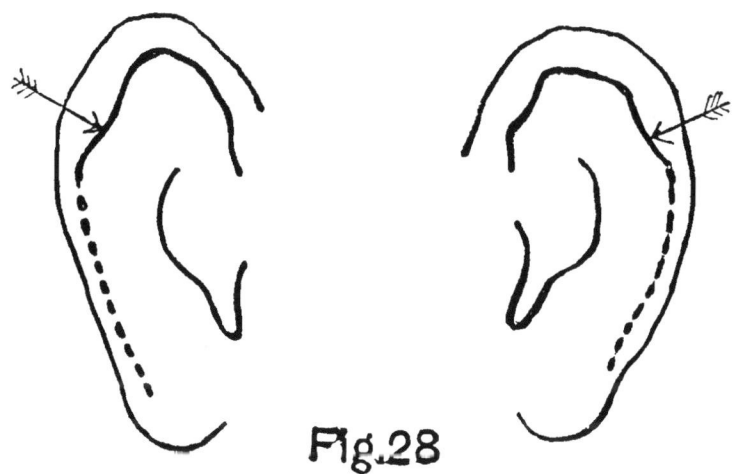
Fig. 28

(4) runs into Division (5). The lobes differ somewhat, but this is owing to unequal piercing for ear-rings.

(Fig. 28).—Here the ears form a tolerable pair, and are also an example of an unusual kind of *helix*. The right ear is somewhat similar to the left, but the *pinna* is more pointed. In the left ear Division (1) is large and bulged, Division (2) runs into Division

(3), whilst Division (4) is bulged and pulled up too high, and it has the knot that indicates a speciality in its use. Division (5) is long and tapered, and the dotted line shows where the delicate indent runs which marks the roll-form of *helix*, giving a tendency to obey laws and make rules for the well-being of others.

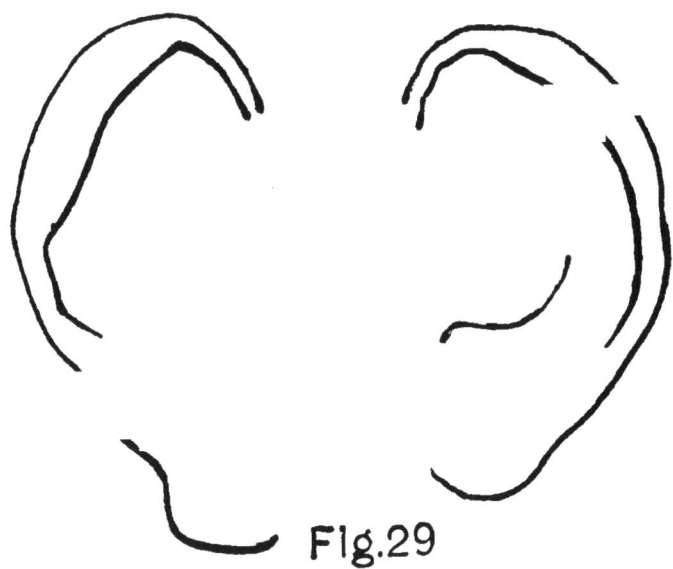

Fig. 29

The orifice is of a very peculiar shape, sometimes found with a taste for music or singing, but not with a general quickness of hearing for other sounds.

In Fig. 29 we have a splendid form of *helix* for the active use of Divisions (2), (3), and (4). They are all joined together into a large bulged rim, Division (4) coming well tapered down to its proper place, with Division (5) distinctly beginning in the roll-shape and then spreading out wide and flat. This would indicate a discoverer in a field of science, for it begins by a strong prejudice—Division (2) —in favour of out-of-the-way subjects— Division (3)—well thought over—Division (4), of which the laws are sought through attention to masses of facts—Division (5).

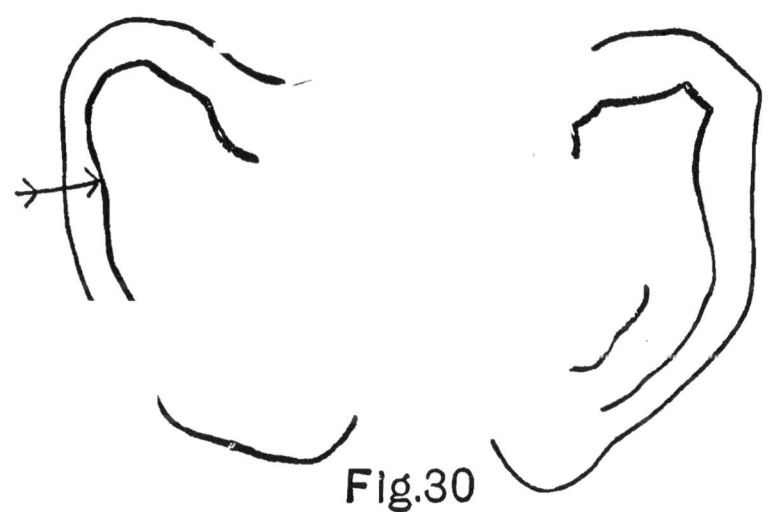
Fig. 30

The rims being so equal in size and shape allow of no intermission of the bent of mind. They belong to Sir John Burdon Sanderson, M.D., Regius Professor of Medicine at Oxford, far-famed in science.

(Fig. 30).—Here we have a most extraordinary example of the Five Divisions in the left ear, all in their own places, and all well proportioned and developed; whilst in the right ear they are again indicated, though with much less precision, but, in addition, there is the knot in Division (4)—indicated by the arrow—which gives a speciality to which the attention is mainly directed. The orifice is of a pear-like shape, in both ears alike, though here not clearly shown, owing to omission of the outline. That these ears must belong to a philologist of the highest rank is inevitable. For Division (3) is of the accentuated shape that is nicked inside at the two ends and with the elbow-form outside, whilst the orifice gives minute attention to the sounds of words, and the roll-form of Division (5) causes the seeking of laws in these subjects. They belong to Professor **A. H. Sayce**, Professor of **Assyriology** at Oxford. It would not be possible to find a more brilliant example of a special character formed for the particular purpose of amassing curious and invaluable lore on ancient languages.

I had already seen all the Five Divisions separately before I found any ears with all of them shown at once in any separate forms.

# FURTHER EXAMPLES

Usually the Five Divisions are to be found coalescing in groups. It is one of the rarer types of ears; and this is the only case in which I have seen each division exactly in place nicked and shaped as if on purpose to exhibit them in proper order.

A peculiar form of *helix* is in the ears of the novelist, Charles Dickens (d. 1870), which have been obtained from two different

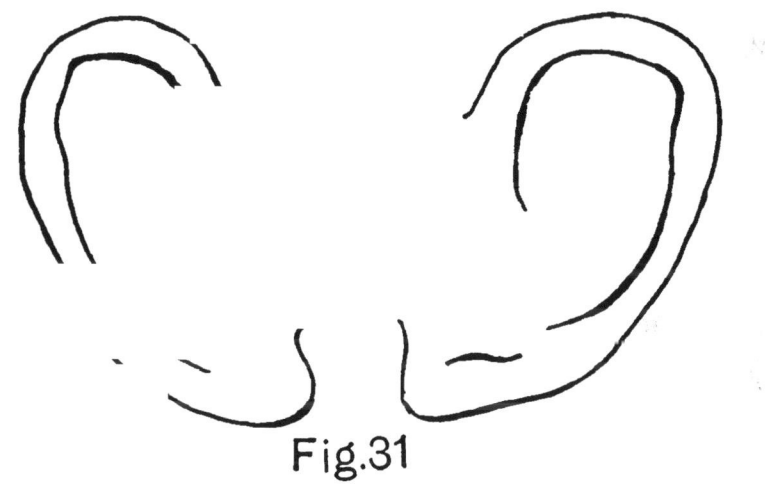

Fig. 31

sources (Fig. 31). The right ear is reproduced by special permission from a photograph by the London Stereoscopic Co., Ltd., 106 and 108 Regent Street, W., and 54 Cheapside, E.C. The left ear, also reproduced by special permission, is from a photograph belonging to Mr. G. L. Lea, Photographic

Printer, St. Alban's Road, Watford, Herts. The right ear has Division (1) joined to Division (2)  There is no Division (3). Division (4) is of a very unusual shape  It is long and tapered top and bottom, with a slight tendency to begin the tapering too soon, and then to give it up and to begin again lower down. If this is to be translated like the rest, it would mean that his powers of attention habitually tried to leave off in the middle of a subject. And if we turn to his MSS. and examine them, the peculiar way in which they are almost doubled in length by additions and insertion between the lines would seem to bear out the suggestion of the rim as to the kind of action of the attention of the mind. It is as if he caught up the train of thought again, and completed it on revision. Division (5) is tapered above and below; this form means dealing *in detail* with masses of people, certainly an absolute necessity to a novelist of his kind.

The left ear is nicked where Division (3) should be, and the part is pushed outwards with a square form, showing an abundant flow of words. Division (4) is long, and tapered above and below, showing good capability of attention. Division (5) is large, and only tapered in the upper part, which indicates dealing with masses of people *as a whole,* and with a novelist these masses no doubt would be his readers that he wished to sway rather than the characters of his fancy, who were already amply provided for by the shape of his other ear.

We do not analyse criminals' ears, as this book deals with the " non-criminal " classes. But their classification can be easily secured by the methods explained in the third chapter, after which the same rules of this new sort of physiognomy can be applied, hinting the causes that led to their devious ways, and perhaps indicating the qualities that may be turned to improvement. Our claim for the somewhat limited but useful physiognomy of the ear is that it is based on observation and scientific forms of identification.

## CHAPTER XI

### HEREDITY AS SHOWN IN EARS

THE question how far heredity can be shown in the ear is extremely interesting when we consider its full import. For by comparing careful portraits of ears with those of the parents, the children may be identified as their offspring when the other features do not give sufficient clue. It is in the recombination of the Five Divisions of the *helix* that the child asserts its own identity, but the kind of divisions are generally hinted at in those of the parents. In the case of the orifice this is equally noticeable, as we shall see later on.

Our first examples are those of a father and mother and two of their sons (Figs. 32, 33, 34, 35).

In Fig. 34 in the *left* ear Division (1) is thick like the father's, but the rest of the

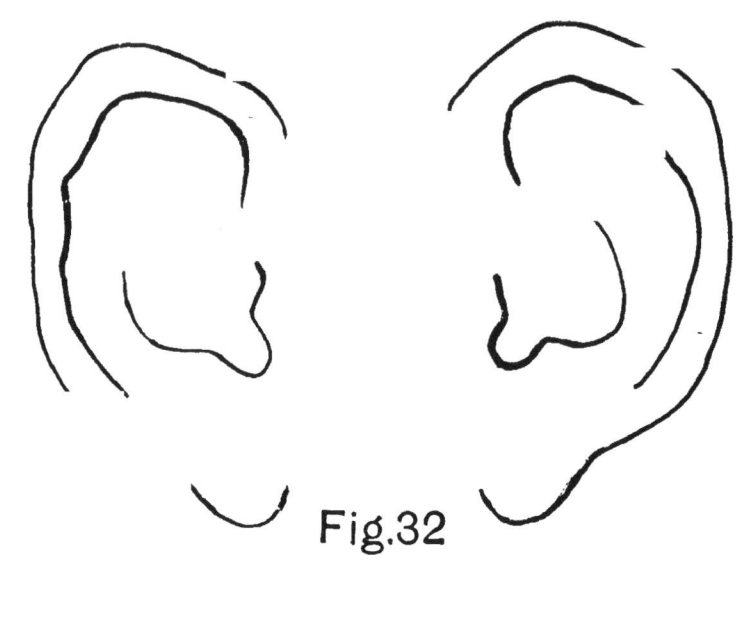

Fig. 32

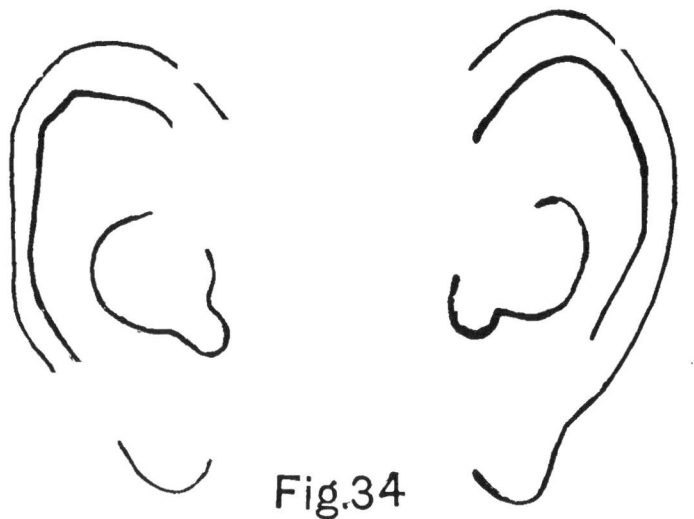

Fig. 34

*Facing page 166.*

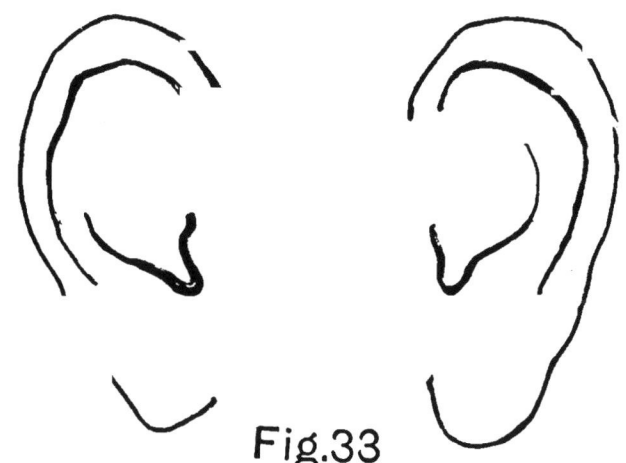
Fig.33

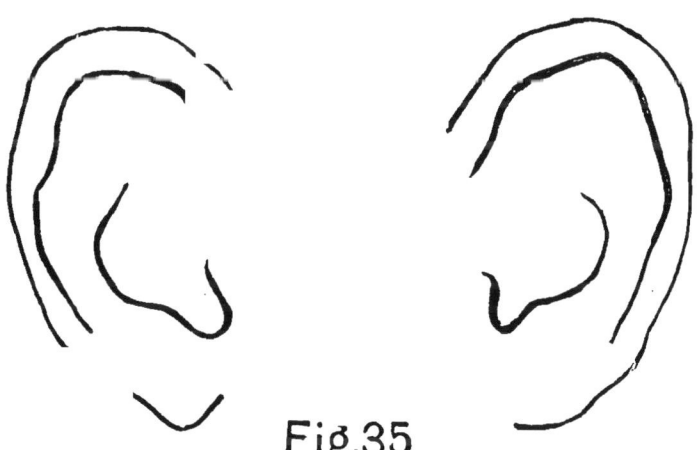
Fig.35

*helix* is narrower, and the lobe is pointed and small. The orifice is like the father's, but the whole *pinna* is narrower. We may here observe that the father's ear is very wide, and the mother's ear is rather narrow. The *right* ear of Fig. 34 begins like the father's *left* ear (Fig. 32), Division (1) running into Division (2), and both these divisions are thick. Division (3) has an original shape of its own, not borrowed from either parent, and the *helix* is again narrower than the father's, ending with a small pointed lobe. The orifice of the *right* ear is like that of the father's *left* ear.

In Fig. 35 the *left* ear is like the father's *left* ear (Fig. 32) on a smaller scale, with the divisions less marked, but Division (2) is pushed up where it joins Division (3), and the latter runs into Division (4) without a nick. The top of the ear is flat, like the mother's (Fig. 33). The orifice is in shape between that of each parent. In the *right* ear the resemblance to the father's *right* ear is noticeable, but all on a smaller scale, and with the difference of having a flat top to this ear also, again like the mother's ear.

We have already given the ears of Sir John Stainer, Mus. Doc., lately Professor of Music at Oxford, the well-known composer (Fig. 16, chap. ix.).

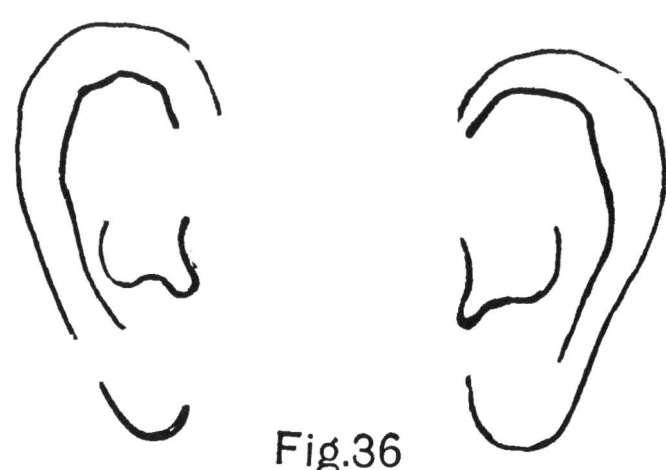

Fig. 36

These of his wife, Lady Stainer, and of

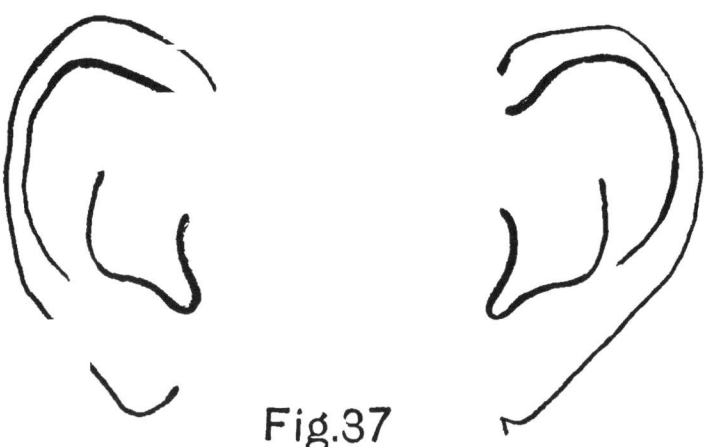

Fig. 37

their eldest son, Mr. J. F. R. Stainer, a barrister, are shown in Figs. 36 and 37. A

# HEREDITY SHOWN IN EARS 169

peculiarity of heredity in orifices of ears is illustrated in Fig. 37. The *left* orifice is like that of the mother's *left* ear (Fig. 36), and the *right* orifice is like that of the father's *right* ear (Fig. 16, chap. ix.). It is often found that a child reproduces one orifice from each parent's ear, but not always in the same ear. In Fig. 37 the *helix* of both ears has a general likeness to that of the father, but it is much narrower, and the lobes are shorter, the right lobe being pointed like the mother's.

We are fortunate in being able to give an example of a complete family in these seven specimens. The parents' ears come first. The five children range from eighteen to six years of age, and are placed in order of age, beginning with the eldest.

At first sight they seem to be all of the same size, and it is probable that only the two last pairs, belonging to those considerably under fourteen years of age, are ever likely to grow larger.

The thick *helix* of the father (Fig. 38) reappears in all the children. The orifice of the father's *left* ear affects the *left* ears of Figs. 40, 41, 42, and 44, modified (except in Fig. 41) by the larger lower inlet of the orifice of the mother's ear. Fig. 43 has the orifice much more rounded in its contours than the father's.

Division (1) is large in all cases, except the *right* ear of Fig. 41.

The peculiar bulged Division (2) of the mother's *left* ear (Fig. 39) reappears in *both* ears of Fig. 44.

The way Division (3) and Division (4) are run together in the father's *right* ear (Fig. 38) reappears in the *left* ears of Figs. 40, 41, and 42, whilst it occurs in *both* ears of Fig. 44 in a lengthened form.

Division (5) is well developed in both parents, but of different shapes. Both shapes are adopted and modified by the five children, the tendency being chiefly to follow the wider outline of the mother's. Fig. 44, however, very nearly reproduces the outlines of the father's ears at Division (5), although the whole *right* lobe is wide like the mother's *right* lobe.

None of the children have the knot in the mother's *right* ear (Fig. 39). In this respect they all follow the father, who has none.

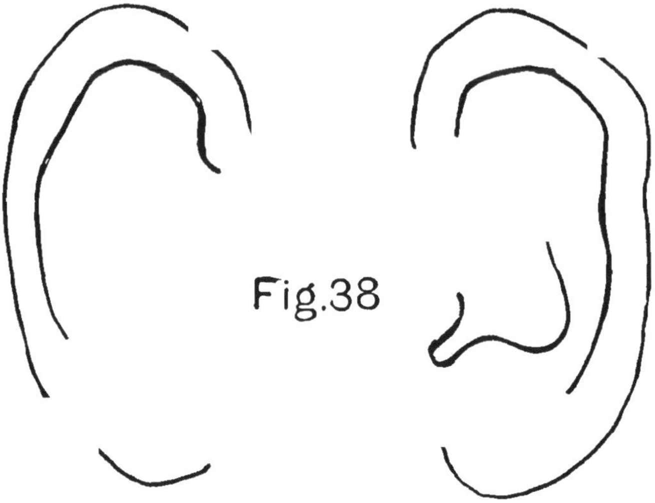

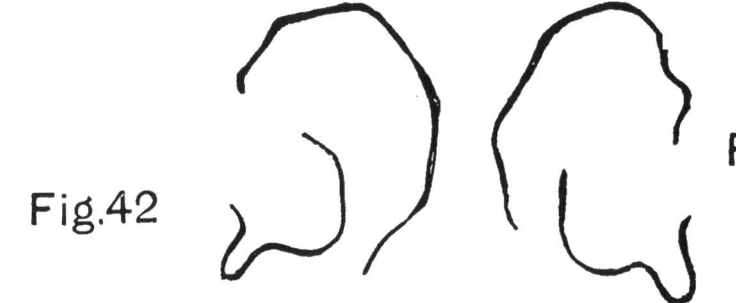

Fig. 38

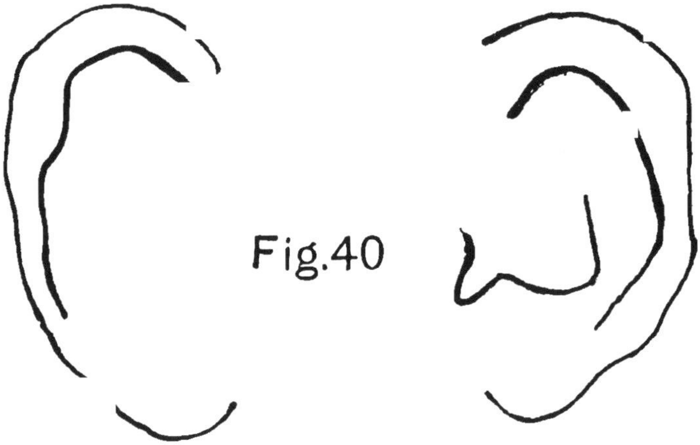

Fig. 40

Fig. 42

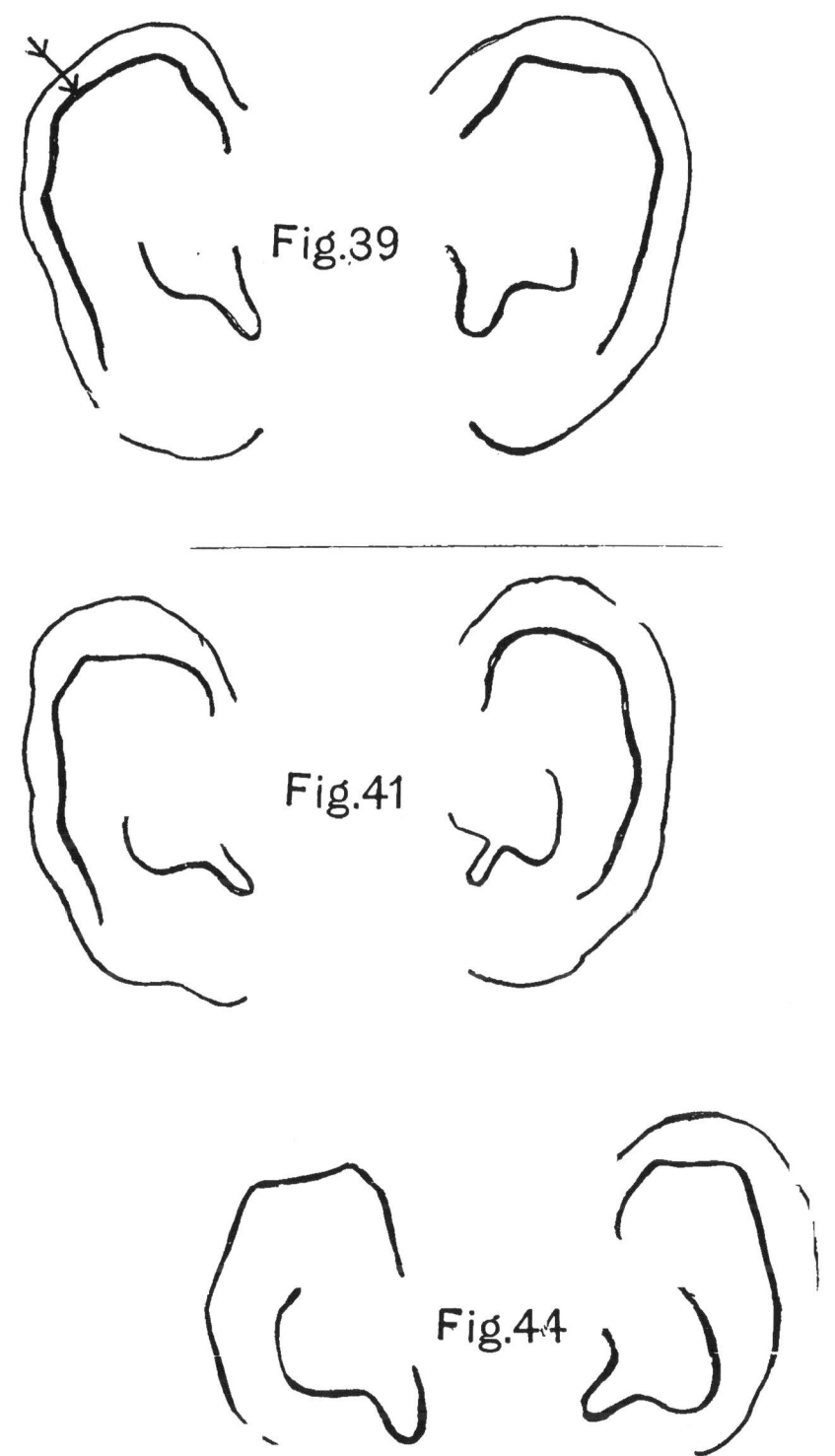

# HEREDITY SHOWN IN EARS 171

The strongly-marked nick where Division (1) joins Division (2) in the mother's *left* ear (Fig. 39) is hinted at in the *left* ear of Fig. 42, and it also reappears in *both* ears of Fig. 44. The nick at the other end of Division (2) in the mother's *left* ear (Fig. 39), where it marks the place where Division (3) runs into Division (4) and pulls it too high, is hinted at in the *right* ear of Fig. 41 and in the *left* ear of Fig. 42, and it also reappears well marked in *both* ears of Fig. 44.

The pear-shaped orifice in the mother's *right* ear (Fig. 39) does not seem to reappear in the children. In Fig. 40 and Fig. 42 the *right* ear orifices have not been outlined, but other indications infer that in each case these resemble their own *left* ear orifices, as also in the case of the father's ears (Fig. 38).

Another point to observe in heredity is the shape of the top of the *pinna*. If there is any special curve or flatness in the top of either ear of either parent, it is generally reproduced in the opposite ear of one of the children. As far as we have observed, the chances are equal as to whether it is the ear of a son or a daughter that may take after the ear of the father or the mother. In the family at present under consideration, the father's ears (Fig. 38) have both a more complete arch than the mother's (Fig. 39), whose *left* ear has a tendency to the "elbow" form of *helix* just where Division (2) runs into the short and thick Division (3), which latter pulls Division (4) upwards out of place. The eldest daughter (Fig. 40) has a tendency to straightness at the top of *both* ears, evidently got from some ancestor, as it is not seen in either parent. The second daughter (Fig. 41) has her *left* ear arched like her father's *left* ear, whilst her *right* ear is drawn out at the top somewhat like her mother's *left* ear. The eldest son (Fig. 42) has the *left* ear topped like the *left* ears of both his father and his mother, *i.e.* the outline is like his father's, only it is widened like his mother's, and the arch is lowered, whilst his *right* ear is topped almost the same as his own *left* ear. The second son (Fig. 43) has the gentle arch in both ears like that in his father's *right* ear, only that the whole top being wider it slopes off more gradually. The third son (Fig. 44) has the outline of the top of the *right pinna*

# HEREDITY SHOWN IN EARS 173

like that of his mother's *left* ear, and the *left* one also like her *left* ear. We have already remarked on the *helix* of this very hereditary and yet entirely original ear.

As the *lobe* follows the shape of the chin, it is apt to vary in the growing ear, nor does it exactly fit the growing chin, but it seems to be fully developed earlier than the chin. We may notice in this family there is very little direct reproduction of the parents' lobes amongst those of the children. But the rather square and wide form of the lobe in the mother's *right* ear (Fig. 39) is somewhat followed in the *right* lobe of the eldest daughter (Fig. 40), and in the *right* lobe also of the youngest son (Fig. 44). In each case, however, the distinct "elbow" form of the bottom of Division (5) as it reaches the lobe in the father's *right* ear (Fig. 38) is added above the mother's shape of lobe in Fig. 40 and Fig. 44, and the tip of the *left* lobe of the eldest daughter and the youngest son resembles the *right* lobe of the father's ear.

It is a fact not generally known that the profile of the chin differs on each side, even if the chin appears to be perfectly balanced in the full-face view. Yet where the lobes differ very much from one another, it is usually an

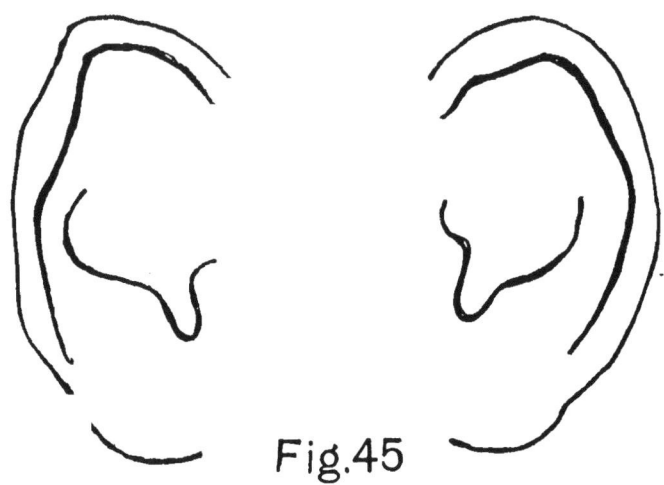
Fig. 45

indication that some chance has deformed one of them, and the chin on both sides, seen in

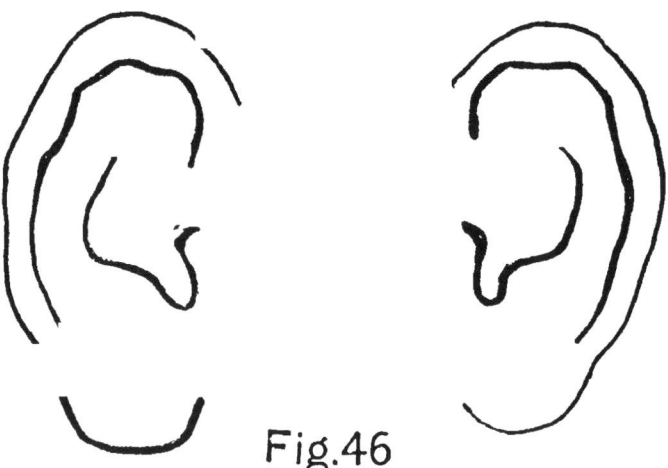
Fig. 46

profile, will be more like one ear than the other. We cannot be sure from this shape

# HEREDITY SHOWN IN EARS 175

of the ear of a boy of six, as to which lobe his chin will follow in shape. As a rule, however, the chin grows up to match the lobe of the *left* ear. We may perhaps expect this chin to become rather flat with a little point in the centre that will not show so much in the profile as in the three-quarter's view.

The ears of Dr. Richard Garnett, late of the British Museum (Fig. 45), and of his

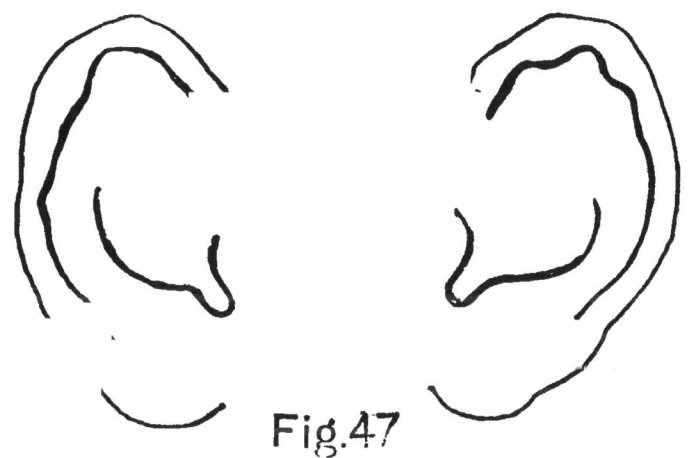

Fig. 47

wife (Fig. 46), and of their eldest daughter, Mrs. Guy Hall (Fig. 47), show in a marked manner the tendency of a child to select an ear-shape from each parent.

The father's *right* ear (Fig. 45) is copied in the *right* ear of the daughter (Fig. 47) with but slight modifications, whilst the daughter's *left* ear (Fig. 47) copies the mother's *left* ear

(Fig. 46), except that in the shape of Division (2) it follows the mother's *right* ear. The orifices of both ears of the daughter (Fig. 47) resemble that of the father's *left* ear (Fig. 45). The lobes of the daughter's ears are smaller and are most like the father's.

The way in which children's ears modify their selections from their parents' ears can be seen in the following examples :—

Division (1) is large in both parents (Fig. 48 and Fig. 49). It is not so thick in the daughter's ears (Fig 50), or in the son's (Fig. 51), but the length of Division (1) in Fig. 50 is like the father's, and in Fig. 51 it resembles the mother's.

Division (2) in the father's *left* ear (Fig. 48) is like Division (2) in the daughter's *left* ear (Fig. 50), and in the son's *right* ear (Fig. 51), but in the other ears Division (2) resembles neither parents' form.

Division (3) is small and yet bulged in the father's *right* ear (Fig. 48), and it reappears large and thick in the son's *left* ear (Fig. 51), and in his *right* ear Division (3) runs into Division (4) and drags it too high, whilst Division (4) ends with a nick where a narrow,

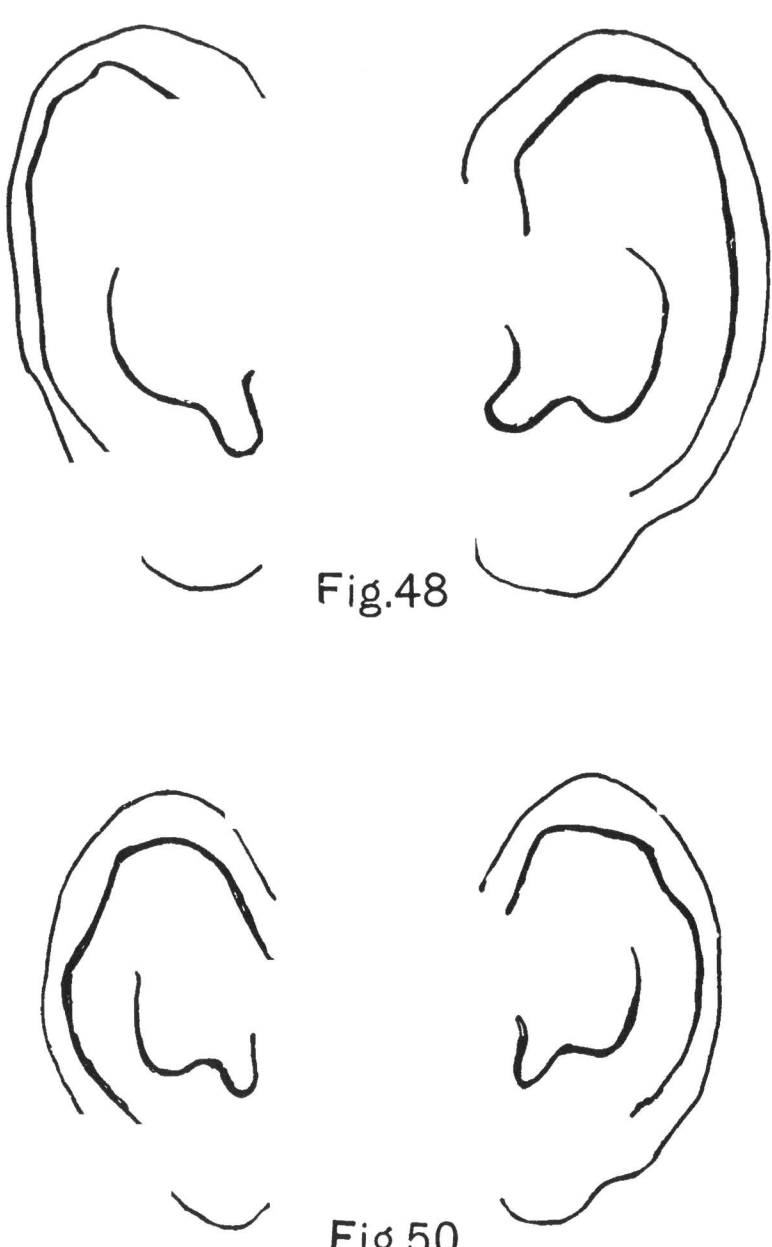

Facing page 176.

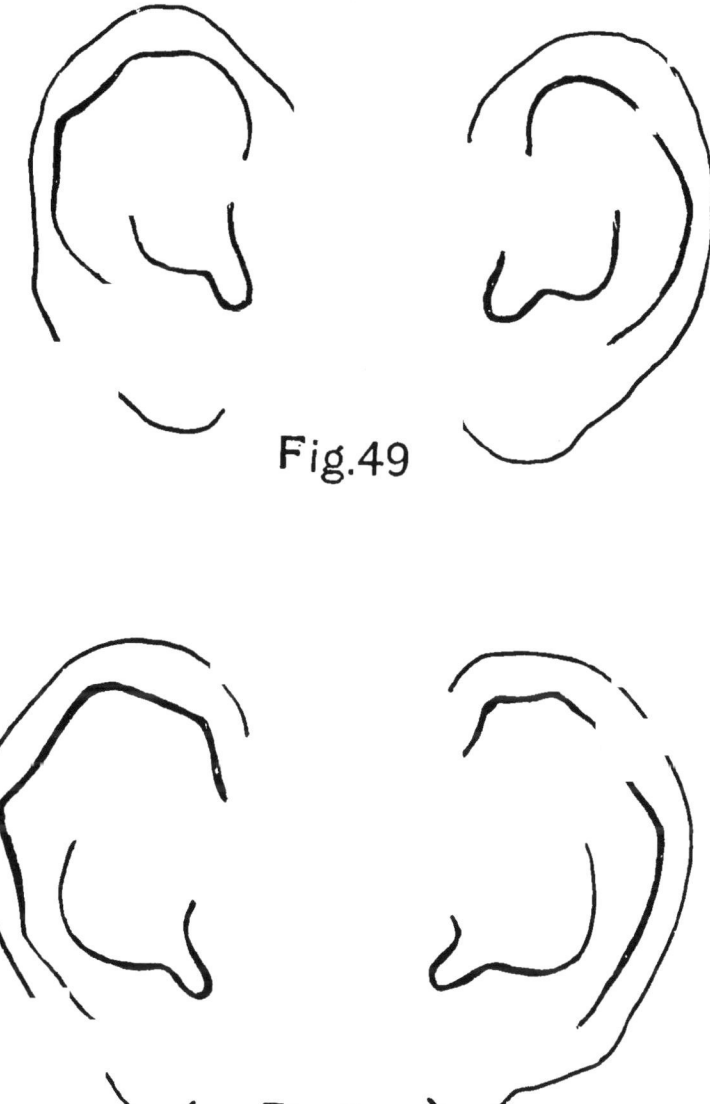

Fig.49

Fig 51

# HEREDITY SHOWN IN EARS 177

straight Division (5) begins, just like in the mother's *right* ear. The ears of the daughter (Fig. 50) have not got Division (3), and Division (4) is pulled rather high, and is therefore shorter than the father's, although longer than the mother's.

Division (5) in the daughter's *left* ear (Fig. 50) resembles the father's *left* ear (Fig. 48), but in the *right* ear it is like the mother's *left* ear (Fig. 49). Division (5) in the son's ear (Fig. 51) follows the father's *right* and *left* ear respectively.

The tops of the *pinna* all vary, except the *left* ear of the daughter (Fig. 50), which takes after the *left* ear of the father (Fig. 48).

The orifices of the daughter's ears (Fig. 50) resemble those of neither parent. Those of the son (Fig. 51) are somewhat like the father's in width.

The lobes of the ears of both son and daughter are smaller and more pointed than those of the parents.

It will be noticed that the remarkably long Division (4) in the father's ear (Fig. 48) has not been copied by either of the children.

This particular form of Division (4) is generally individual rather than hereditary.

The above extremely interesting varieties of

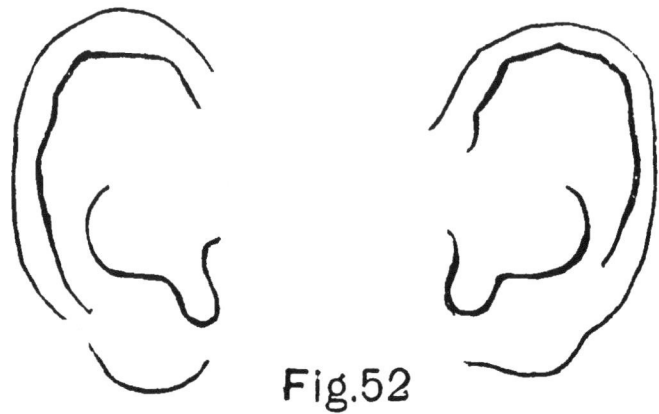
Fig. 52

ears in members of the same family belong to Dr. Griffiths, lately Chief Surgeon of the Royal

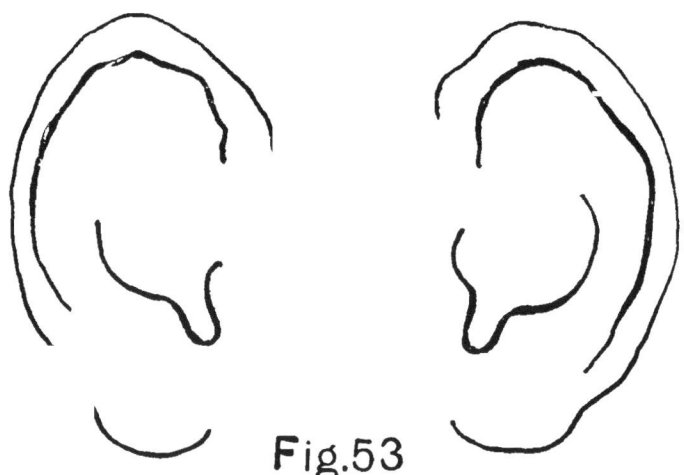
Fig. 53

Naval Hospital, Haslar, with his wife, their only daughter, and youngest son.

That the ears of twins should differ, mani-

# HEREDITY SHOWN IN EARS 179

festly comes under the heading of Heredity. In these examples of the ears of twin-sisters, the ears differ so much in size and in shape that identification is easy. It happens that these twins are not alike in feature, and this can be gathered from the lobes indicating a different size of chin for each, the varying length of the *pinna* implying also a different length of the nose for each.

A case of twin-sisters, who closely resemble each other in appearance, has come under our notice, but we are not permitted to give the ears. We may mention that these ears vary also, both as to size of *pinna* and shape of *helix*. They are not quite as dissimilar as the above, but the *helix* varies sufficiently for instant identification.

## CHAPTER XII

### EARS AS PORTRAYED IN SCULPTURE AND PAINTING

FROM the sculptures and bas-reliefs that remain to us, belonging to the Assyrian period, we find that the Assyrians preferred to carve a very round ear, almost circular at the top and runing down suddenly to the lobe, from what we may now refer to as "the middle of Division (4)" in its normal situation. Many specimens are to be seen in the slabs let into the walls along the Assyrian Galleries at the British Museum. This form of ear is repeated regularly on all profiles, whether of kings or of captains or of common soldiers, with the singular exception of some of the captives or slaves—showing that the sculptor knew what he was about. The circular shape is also used for the Assyrian human-headed bulls at the entrance to the Egyptian Hall at the British

Museum, except where bulls' ears have been employed. Since we can see for ourselves at the present day that no nation or race has ears made after one unyielding pattern, in spite of distortions from ear-piercing as practised amongst savages, we may suppose that this round ear was a conventional form, intended to indicate an Assyrian ear in the same formal way as curious spirals are made to do duty for the curled Assyrian beard. Very probably some early king had this form of ear. Or perhaps a master-sculptor sat for his own ear, reflected in an early silver mirror, and he imposed the shape as a tradition. Or it may be that the Assyrian hand held the chiselling tool in a manner that made a kind of semi-circular ear, to be done more surely and more gracefully than any other shape. This much can be said in its favour, that it has a supremely decorative effect in these weird historical documents in stone, harmonising with the rounded aquiline of the nose and with the general tendency to circles in the rings of the hair and beard, and contrasting with the sharp, straight lines of arrow-headed inscriptions carried across the lower part of the figures in the cuneiform character, in itself the most artistically decorative alphabet ever invented.

No other nation has seen fit to adopt this form in its portraits, though the Egyptians, with whom the writers of cuneiform have of late been shown to have been in constant correspondence, have also a tendency to carve the tops of the ears in a somewhat too arched and decorative way, elongating the whole *pinna* at the same time. The Egyptians delighted in giving the natural ears of their animal-headed idols, which shows their knowledge of the subject. They possessed, however, the artistic perception that the human ear, being by nature beautiful and intricate in shape, would attract too much attention if represented with the same amount of finish bestowed upon the other features, and accordingly it is kept in subordination. They invented a peculiar trick of putting the orifice in the wrong place, *i.e.* too high, and formed like a mere indent, which deludes the eye on a cursory glance, and actually induces it to return at once to the consideration of the more accurate profile  The outline is sometimes correctly given in relation to the jaw, but while the bas-relief of the *pinna* is purposely scamped, the exquisite carving of the cheek and chin shows that the sculptor must have had an eye for form and a hand for execution which could have caused him to represent the ear faithfully if he had so chosen. We are driven otherwise to suppose that the royal owners had very sorry specimens of ears, and that attention was not particularly desired as to their *helix* and orifices.

The Egyptian ears are usually placed vertically against the side of the head. It is not possible that every Egyptian should have had them in this position. Possibly it was considered to be a beautiful and kingly attitude  And, indeed, as ears follow the slope of the nose, and the Egyptian nose took the flat form of the aquiline, probably a good many ears were tolerably bolt upright. Although the ears are often sculptured in right proportion to the nose, yet they are sometimes hoisted so high that the bottom of the lobe is scarcely below the lower eyelid. This gives a new decorative effect of an extraordinary kind, which must have been intentional. It alters the expression of the rest of the features, and by preventing them from looking too human, subtly suggests that the statue of the king represents something superhuman. I have not observed this position of the ears except in statues of Egyptian kings and gods, or on the carved coffin lids, where a decorative and superhuman effect is the first object. Nevertheless, there is an exception, and Tahutmes IV. (B.C. 1414) is remarkable for having a really human ear, with the orifice in the right place and the *helix* and lobe distinctly carved as if representing a true portrait of an ear. In spite of the careful carving, it is much too large for the face, and does not tally in outline with it. But the head-dress that closely fits the forehead, and then stands straight out behind the ear at right angles to the head, seems to demand some huge ornament to be set in the cavity to relieve its austere lines and to set off the small and delicate face. It was a bold sculptor who ventured to enlarge an ear for the sake of decoration. We are grateful to him, for the ear is essentially of present-day type, showing that ears do not belong to time or race The head is slightly uplifted, allowing the slope of the nose to be seen very clearly, but the ear is placed vertically against the side of the head, so that the front of the *pinna* is too near the cheek by exactly its whole width. It does not appear to have been a "restored" ear. In the gigantic heads at the British Museum, the granite ears on Amenophis III. are rather sketchy in form, whilst those of Rameses II. are huge and roughly shaped. In a *stele* of Pai (see *Lanzone*, Plate CCCXL.) four ears are carved in bas-relief to show the god invoked hearkened to all things. The upper two are placed one above the other, the lower two are placed side by side, and in the space between there are two right eyes, one above the other. All the ears are right ears. The *helix* is cleverly carved and varies a little in each, but the *pinna* is elongated by enlarging the lobe in a conventional way till it forms about one-third of the ear.

Among the Greeks and Romans the sculptor carried out the artistic feeling of subordination in the ears of their statues, except in portrait busts, although they did so in a different manner to the Egyptians. The hair, being worn thick and waving, helped to shade the ear more or less when desired, and the type of ear chosen for statues was of the elementary kind, where the *helix* was gently rounded to harmonise with the rounded cheek instead of with the outline of the jaw-bone, and the lobe was not large, whatever might be the size of the chin. The hollow of the ear, though in approximately the right place, merely suggested the shape it would assume to a casual glance from a distance. That this was done on purpose, from a conventional rendering of a Grecian idea of beauty, is proved by the greater care with which the ears of portrait busts are rendered The short hair of the emperors, senators, and generals compelled their ears to be a visible part of their recognisable features, and the way in which the outline of the *helix* repeats the outline of the jaw-bone proves the genuine representation of the original ear. Julius Cæsar, Augustus, and Cicero are examples. The ears of Marcus Aurelius are well done in the best busts, but his head was copied and re-copied frequently, till the ears were scarcely the same, and later on it is evident that some alien model was substituted for them.

The best carved bust of Julius Cæsar corresponds with the best effigy of him on a coin, as regards the ear. By comparing other likenesses of him, we see how the original form of ear has been altered by a careless or a less skilful hand in reproduction or in copying. The original form was of a thoroughly well-developed type, following the outline of the jaw and chin and the slope and length of the nose, and with all the Five Divisions separately marked in the *helix*. They are, in fact, so well carved that we can read their physiognomy, which certainly is unique, and could belong to none but Julius Cæsar. As he was known to be particular about his personal appearance, it is not unlikely that he insisted on his ears being as accurately rendered as his other features; he could not but be aware that they were remarkably shapely ears.

Although a conventional ear was adopted for statues, occasionally a real ear was substituted, giving a startling air of superabundance of life without any apparently adequate cause. We may refer our readers to the statue of

Hermes with the infant Dionysos, by Praxiteles, which has a very beautiful ear. The proportion is true, the length being twice the width. The *helix* is gently curved and tapered, whilst the inclination to mark each of the Five Divisions by a little nick inside and a slight " elbow " form outside, is very noticeable, because so rare. The lobe is well curved and pendent and of medium size. The orifice is large, slightly square at the bottom and then slanting up to fit the slope of the ear. The artistic physiognomy of this ear points to it as belonging to Praxiteles himself, for assuredly he could have got no other model for it.

The sculpture and the bronzes of the Middle Ages appear to adopt ears by chance, and for centuries no feature was so inadequately represented. Sometimes they seem to be dropping off, or again to have been stuck into the skull with a thud that would have gone through it in a real head. Or they are made too small and placed too high, or brought forward into the cheek—especially in women's faces, where a small ear close to the nose was considered exquisitely enchanting.

In modern times there has been some endeavour to represent the ear in sculpture in a more adequate fashion. Yet portrait busts seldom even now give the genuine ear. Perhaps it may be too large for the sculptor's fancy although handsome in shape, or it may be too complicated in its folds for easy reproduction. The annual show of marvellous ears in the Royal Academy is a proof that the subject has not been sufficiently studied, for otherwise the originals of the portraits would anxiously desire that their own ears should be represented with as much care as the eyes and nose. A word must be said of the cheap little portrait busts of celebrated men sold in small shops. The ears might be meant for handles by which to lift these heads, rather than for human features.

By observing the way ears have been represented in oil paintings and frescoes amongst the chief productions of the best artists of ancient and modern times, we are compelled to acknowledge that this feature has received hard usage at the hands of those who should have been more kind, inasmuch as they must have known better. No one, except the otomorphologist, can more fully appreciate the beauty and the manifold varieties of the ear than the artist. The colour and delicacy of the skin attracts attention as well as the beautiful curves of the *pinna*, the *helix*, and the *concha*, besides the varied forms of the *tragus* and the *anti-tragus* that seem to stand to challenge each sound before it is permitted to enter the ever-open door. The ear is extremely difficult to draw. And several of the great artists seem to have adopted an ear-form of their own, which becomes one of the signs by which a doubtful picture can be authenticated. With regard to this point, we would refer to the *Life of Lotto*,[1] the Venetian painter, where this is frequently mentioned, and a further account promised for the future. I have observed that the greatest artists drew men's ears like those in their own portraits whenever they could, and that they took some favourite model's ears for their feminine heads. This can be tested by noticing the outline of ear and lobe, and whether they coincide with the jaw-bone and

---

[1] *Lorenzo Lotto : An Essay in Constructive Art Criticism*, by Bernhard Berenson, 1895.

chin, and whether the ear is in proportion to the nose and has the same slope.

Raffaelle used to draw ears that must have belonged to the face he was painting; they tally exactly to the face and are equally accurate in form. For example, notice the foreground figure of a kneeling woman in the great picture of "The Transfiguration." She is in profile and her ear is carefully given, it tallies in every respect with the face, and both are beautiful. It is the arrangement of the hair that prevents too much attention being absorbed by the ear, which is therefore kept in artistic subordination although it is perfectly drawn. It was also by the hair that Praxiteles subordinated the interest of the splendidly human ear he gave to his Hermes with the infant Dionysos, already mentioned.

The tendency of the inferior artists is to draw ears that have no *anti-tragus*, with a rim like a hem all round, a heavy lobe that appears to grow out of nothing, and a dark and shapeless splash of paint that does duty for *anti-helix* and orifice in one. No nation or period can claim the shapeless ear for its own speciality, for the inability to draw an ear besets the artists of every country and of every time. In portraits the ear is frequently shirked by means of overhanging headgear or hair. There was a period of about eighty years when short wigs or long hair hid all but the lower half of the ear, between 1770 and 1850. The peculiar way in which this unfortunate truncated ear is hitched on to the side of the face suggests that it was fastened to the wig and put off and on daily. Profile portraits were common at the time, for the style of wig suited the profile better than the full face. The black *silhouettes* helped to perpetuate this fashion, and on some of them can still be seen dainty little curves in gold, supposed to represent the ears. Ringlets of ladies were added in much the same manner.

Photography has changed the whole world of portrait painting, and in nothing has it done more good than in the recognition of the rights of the ear. Full-face portraits were chosen at first, but the sticking-out of the ears appalled the beholders. Every way was tried to disguise the size of the ears. Looking through old photograph books, we can see to what straits the photographers were brought.

For if the ear was foreshortened the nose became a blob, and when the nose was in a perfect three-quarter's-face the ear again became conspicuous. Moreover, as photography always diminishes the apparent size of the eyes and increases that of the mouth, profiles went suddenly out of fashion; since a small eye, a huge mouth and chin, a nose of uncertain regularity, and an ear that would insist on being full face, so to say, when all the rest was in profile, made the picture extremely trying as a portrait. No one had known, apparently, how big even the smallest ear could be. "What ears!" is the common exclamation on looking over family treasures of photographs of the different members at all ages. The fashion of showing women's ears came in about 1860, and as the Queen and the Princess of Wales have both fine ears the custom continued for many years. At the Diamond Jubilee, in 1897, a revival of the drooping hair called "Early Victorian" suddenly set in. This fashion will be popular with artists and photographers, as dismissing the problem of what to do with the ear.

Modelled ears in clay, wax, or silver were

placed from days of old in temples and churches as votive offerings in gratitude for recovery from some affection of the ear. They were used by the Romans as well as by sufferers in later times. After reading the ancient cures for deafness, one would hardly expect to find any ancient votive offerings on this score. Nevertheless, several have been found, and those now in the Museum at Bristol were dug out from Roman remains about two or three years ago. They are made of terra-cotta, and from their well-modelled shape nothing seems amiss with them, even the physiognomy can be deciphered; but in each case the orifice is of the somewhat restricted kind which indicates inclination to deafness. Perhaps some temporary attack had occurred or was threatened, and had been cured for the time being. The form does not show hopeless deafness, but a hardness of hearing. The modern votive ears of silver, in the same place, are absolutely conventional, and are merely stamped on thin silver plates.

In the Pitt Rivers Collection in the University Museum at Oxford there is a small case containing votive offerings, including several ears. One terra-cotta tablet has a pair of ears affixed. These two ears differ a little in size, and still more in shape, yet they may possibly belong to the same person, and it does not seem likely that a right and a left ear of different persons should be modelled for one tablet. The orifices are both very small and conventional, one being a round indent in the very middle of the ear, and the other is shaped like an inverted fan. Several of the ancient votive offerings in this case were found by General di Cesnola in the ruined temple of Golgoi during his excavations in Cyprus, but which these are has been left unspecified. In General di Cesnola's book there are woodcuts of some of the ancient votive offerings that he found, and amongst others I have identified this pair of ears (*Cyprus*, by Gen. L. P. di Cesnola, 1877, p. 158). Another terra-cotta right ear on a plaque is long and well-modelled as to the shape and the *helix*, but the orifice is entirely conventional, long, bolt upright, and very narrow.

Among the modern votive ears in the Oxford Collection there is a large one from Brittany, stamped in thin white wax, without a plaque. It has a very thick *helix* at the top and sides, running away to a thread for the lower half of the ear, with a long square lobe and a hole through it, as if for a mock ear-ring. The orifice is large and oddly shaped, as if it had lost its way in the upper part. Evidently this is a conventional offering, and without any definite relation to the owner of the ear that should desire to buy a votive one.

There are also four stamped silver votive ears from Dalmatia, Belgium, Italy, and Greece which deserve attention. They are all modern (1896). That from Spalato is very thin, and is stamped on a plaque. The top of the ear is almost as flat as a ruler, and the *helix* is narrow all round. The ear is wide, yet it runs suddenly inwards to a small pointed lobe. The orifice is almost in the right place, but it is so very small that one can hardly suppose it to be other than a deformity, perhaps caused by the complaint. This looks like a rough portrait of an ear, possibly used as a model for sale, and taken from a chance deaf person.

The ear from Antwerp is carefully struck in a plaque that has a raised dotted background and an edge of large dots. It is a real portrait of an ear, with the physiognomy intact, and the orifice is of a good useful size. The recovery from deafness must have been very complete, if indeed it ever suffered from such an infliction. Probably it was merely a handsome ear that was chosen in order to make an attractive votive offering for sale.

The ear from Naples is carelessly stamped on a plaque, a mere rough *helix* and shapeless orifice, altogether about as unlike an ear as could be.

The ear from Athens is remarkable for having no plaque. It is neatly shaped, and the *helix* is distinct; but the orifice is too low and vague in shape.

# CHAPTER XIII

### CONCERNING EAR-RINGS AND EAR-LORE

IF we examine the ways in which nations seek to adorn themselves, we become full of amazement at the customs of the savages and of admiration at the artistic beauty attained by the civilised communities. Yet amongst both savages and the civilised folk the delight in the ear-ring is a bond of union none the less remarkable for not being recognised. We may call it a barbarous custom to allow the ears to be pierced, but it lingers amongst women, although men have discarded it for themselves.

The origin of personal adornment seems to be a desire not only to be superfine in appearance, but also that it should be attained at a difficulty or cost which will ensure distinction. Some savages are extremely vain of their complicated plaits of hair, and they are obliged to rest their necks on wooden blocks for

# EAR-RINGS AND EAR-LORE 199

pillows in order to avoid disturbing the bunchy mops at night. It must be a very uncomfortable and inconvenient form of repose, a kind of constant rehearsal for a beheading that is neither ordered nor expected. Yet the blocks also are matters of pride, and are plain or inlaid, according to taste. Savages have another way of making discomfort a proud distinction, by slitting the lobes of the ears and inserting tufts of grass or leaves as ornaments. Others put in discs. Lobes thus distended must be somewhat precarious treasures, and may best be described as " of no value to any one but the owner," unless when the discs are replaced by plugs of tobacco.

As no one can possibly tell what made women take to ear-rings when they must have had ample opportunity of seeing how unbecoming they were to the warrior-braves, we may pass at once to the following Mahometan legend:—"Sarah, being jealous of Hagar, declared she would not rest until her hands had been imbrued in her bondmaid's blood. Then Abraham pierced Hagar's ear quickly and drew a ring through it, so that Sarah was able to dip her hand in the blood of Hagar without bringing the latter into danger. From that time it became a custom among women to wear ear-rings" (see Michælis, *Laws of Moses*, 1814, vol. ii. p. 178—*Notes and Queries*, Jan. 3, 1874).

But the custom was already old before the time of Abraham, as we see by the passage in Genesis xxiv. 22, where Abraham, as a matter of course, when he sought Rebecca for Isaac, gave her at the well "a golden earring of half a shekel weight." We quote from a very interesting account of ear-rings in *Notes and Queries* (Nov. 10, 1877), from which also the following particulars on ear-rings in this chapter are taken:—It was from the golden ear-rings worn by the Isrælites after they left Egypt, and of which they had spoiled the Egyptians, that the golden calf was made by Aaron whilst Moses was in the Mount (Exod. xxxii. 34). The Ishmælites wore golden ear-rings even in battle; and Gideon, after beating them, obtained ear-rings to the amount of seventeen hundred shekels of gold (Judges viii. 24-26).

Amongst the Persians both men and women

wore them, being their most costly ornaments. Augustine condemns this: "Persæ mulierum inaures habent, quod hic inhonestum et illicitum est" (*Op*. t. iv. cap. 115). The Persian king Perozes, who fell in a battle against the White Huns, threw away "a pearl of great beauty that he wore in his right ear that none should wear it again, as he rode headlong and saw a deep pit he could not avoid" (Procopius, *De Bello Persico*, lib. i. cap. 4). Ear-rings were also worn by the Medes "as an ornament of distinction" (Agathias, *Hist*. lib. iii. cap. 28). Juvenal recognises a Babylonian by his ear-rings (*Sat*. i. 104)—

> Natus ad Euphratem, molles quod in aure fenestræ
> Arguerint.

(Born by the Euphrates, as the effeminate holes—*lit*. windows—in the ear should prove.)

We have already referred to the passage in Xenophon (*Anab*. iii. cap. 1), where a Lydian was recognised by his pierced lobe although he had no ear-rings; and Dio Chrysostomus implies that amongst Lydians as well as Phrygians "custom of wearing them was restricted to children, but without difference of sex" (cap. 32). This might account for the Lydian not having been found out at once, as the mark of the piercing would have healed up from not being used after childhood. Besides these nations, the Carthaginians and other African colonists wore ear-rings. Moreover, they were used in India, for when Alexander the Great arrived there he found an Indian prince wearing them (Curtius, lib. ix. cap. 1).

The Greeks, who of all nations must have most fully understood the beauty of the unpierced ear, fell a prey to this aggravating custom when they were already well advanced in civilisation. The reason given is that they misunderstood an oracle of Apollo, that declared " if they wished to have good citizens, they were to put what they held most precious into the ears of their children " (Dio Chrys. cap. 32). Greek avarice prevailed over Greek philosophy, and instead of putting words of wisdom into their children's ears, they put literal gold and jewels into the lobes.

In remote Britain the habit spread, but apparently the men refused to be so inconveniently adorned. At least " ear-rings and necklaces and such-like articles of feminine

# EAR-RINGS AND EAR-LORE 203

finery" were classed together, and fell by the Saxon laws to the daughter at the mother's death (Tit. vi. § 6).

Ear-rings became at last so intrinsically valuable that they got to be considered as marks of rank or royalty. Even Plato, the greatest of philosophers, is said to have worn ear-rings, as he was of noble family (Apuleius, *De Habitudine*, lib. i.). Perhaps his philosophy accepted them as inevitable. If we can for a moment imagine the handsome and majestic Plato as a young child, which he must have been at one period of his momentous career, we may be sure the pretty boy questioned the use and abuse of ear-rings as well as of the other facts of the universe.

The Roman Emperor Galba pledged one of his mother's ear-rings to defray all the expenses of a journey from Rome into Lower Germany (Suetonius, *Aulus Vitellius*, cap. 7). Another historical ear-ring was that owned by Cleopatra, from which she took the celebrated pearl which she dissolved and drank at the banquet she gave to Antony (Pliny, *Hist. Nat.* lib. ix. cap. 7).

Often only a single ear-ring was worn, as

if to enhance its value, but it must have given the face a lop-sided air. We have now finished the extracts from the long article on ear-rings in *Notes and Queries* (Nov. 10, 1877), but with regard to the single ear-ring there are two further notices. There is some doubt whether the "single" ear-ring given by Abraham to Rebecca was intended for the ear or for the nose, or whether it was a jewel for the forehead (*Notes and Queries*, Dec. 8, 1877, and Feb. 16, 1878). The last-named ornament would be the most humane and also the most beautiful adornment.

In the matter of ear-lore there are many curious things to consider. There are all sorts of phrases that refer to the ear, such as "to turn a deaf ear," which is to refuse to listen, or to play "by ear," without knowing the notes of music. "In at one ear and out at the other," powerfully describes an inattentive mind to what is heard. To have "no ear" is simply to be devoid of "an ear for music," whilst to be "all ear" is to give one's full attention. To "give one's ears" for anything is to make a considerable sacrifice for the purpose. This refers to the time when ears were cropped as a punishment. To have "itching ears" means anxiety to hear current gossip. To "set people together by the ears," is to set them quarrelling till they pull each other's ears. In Hone's *Year-Book* (Sept. 14, 1859, "French and English Manners in 1831") "pulling the ears" has another signification. "To bite the ear, on the other hand, was anciently an expression of endearment; and it is still so far retained by the French that to pull a man gently by the ear is the most sure token of goodwill. . . . Napoleon I. did it when in high good-humour. . . . Indeed, I have known persons of great respectability pull one by the ear gently in England. . . . Frenchmen also wear ear-rings as did the coxcombs in Shakespeare's time."

The expression "ear-marked" means marked so as to be recognised. The allusion is to marking cattle and sheep on the ear.

The phrase, "Mine ears hast thou bored," means "Thou hast accepted me as thy bond-slave for life." If a Hebrew servant declined to go free after six years' service, the master was to bring him to the door-post and bore his ears through with an awl in token of his voluntary servitude (Exod. xxi. 6).

A misprint in a copy of the Bible in 1810 has caused it to be known as "The Ears to Ear Bible." The text, "Who hath ears to *hear*" (Matt. xiii. 43) was printed by accident "Who hath ears to *ear*."

We have the old English word *earing* for *ploughing* in the following texts:—"And yet there are five years, in the which there shall neither be earing nor harvest" (Gen. xlv. 6); "In earing time and in harvest thou shalt rest" (Exod. xxxiv. 21).

We must refer our readers to Brewer's *Dictionary of Phrase and Fable* for further examples, from which the above have been selected.

Under the heading "Ear" in the great English Dictionary edited by Dr. J. A. H. Murray, we have an exhaustive list of uses of the word, from which a few may be taken. For instance, "To sleep on the right or left ear" is to sleep lying on one side, whilst "To be able to sleep on both ears" indicates one is free from anxiety. "Wine of one ear" is good wine; this is a French idiom of obscure

# EAR-RINGS AND EAR-LORE

origin. "To hear of both ears" is to be impartial, listening to both sides of a question. "At first ear" means on the first hearing. "To have a person's ear" is to have his favourable attention. The odd phrase "ear-kissing" merely means that it is whispered in the ear, not that the ears are rubbed together as the words would seem to imply!

The word *ear* is used for small projections of inanimate things—such as ear of a bell, by which it is hung. Ears of bombs, ear-shaped rings for lifting them by. Ears of a pump, the support of the bolt for the handle or brake. Ears of a cap, the parts that come over the ears; in men's caps they are often called ear-flaps.

"Ear-wise" means after the manner of an ear of corn.

Besides the ear-shell, to which we have referred in the first chapter, there is a bird called the ear-dove, which has two spots of a dark colour, one on each side of the head. The vegetable kingdom is represented by the ear-wort (*Dysophila auricularis*), a plant supposed to be good for curing deafness. A less pleasing live object is the ear-worm or ear-wig. It is not precisely known what is the origin of this name. There is a popular superstition that this insect delights to enter, and even reside in, the human ear, but this has not been proved to be a fact. Of course bees, flies, wasps, or any creeping or flying insect may have a chance of entering the ear, and the remedy of pouring in sweet oil or warm milk is the same as for dislodging any other foreign substance. Another supposition for the origin of the name is that the gauzy wings of ear-wigs are shaped somewhat in the form of the human ear. In *Notes and Queries* (April 12, 1855) it is erroneously traced to two Anglo-Saxon words, meaning *ear* of corn or *bud* of flower, and *dwelling*. Dr. Murray says that etymology shows that this derivation is not possible.

The following charade on the ear-wig is ingenious, if not strictly accurate :—

### CHARADE. (EAR-WIG)

My *first*, if lost, is a disgrace
   Unless misfortune bear the blame;
My *second*, though it can't efface
   The dreadful loss, yet hides the shame.

# EAR-RINGS AND EAR-LORE

My *whole* has life and breathes the air,
    Delights in softness and repose;
Oft, when unseen, attends the fair,
    And lives on honey, and the rose.
        *Notes and Queries*, November 15, 1851.

Our old friend "Fine-ear" must not be forgotten. He was one of the seven attendants on Fortunio, in Comtesse d'Aulnoy's *Fairy Tales*, as long ago as 1682; he was also in Grimm's *Goblin's Tale*. He could hear the grass grow and even the wool grow on a sheep's back. In fact, he was a microphone in disguise.

# CHAPTER XIV

### THE EAR IN LITERATURE AND SCIENCE

THE ear has been well treated on the whole by writers in prose and poetry. Its beauty is unquestioned in fiction, and a heroine has always dainty ears. Yet this vagueness of description leaves much to be desired, and from no source can we gather what it is like without the aid of an illustration from a photograph. Drawing, painting, and sculpture are often in vain. Where artists — the men of trained observation of form — have been uncertain, how could poets "in a fine frenzy" be more accurate? Ears are described as being like a "shell"—that is, like a hard, shiny, unsympathetic substance of an unnatural white or salmon-pink colour. Whereas part of the charm of an ear lies in the marvellously fine skin that covers it, of admirable texture, and of delicate pink or living white

# IN LITERATURE & SCIENCE

in tint. An ear can flush, can turn purple, can seem puffed up or abased; in short, it has a number of powers and qualities in common with the face. It has also one special quality in itself that draws and attracts with a magic charm, we know that it will gather in our words and speed them swifter than a lightning flash to the very citadel of the soul.

Perhaps we may know that an ear is large or small, but to a poet an ear is always small. In the following passages from Shakespeare it is to the use of the ear rather than to its appearance that reference is made, which shows how little the ear was observed for its own sake. He gives us incidentally a hint that monsters were hard of hearing—

> O, 'twas a din to fright a monster's ear!
> *Tempest*, ii. 1.

In contradistinction to this, he puts the lover's ear before that of the musician—

> A lover's ear will hear the lowest sound.
> *Love's Labour's Lost*, iv. 3.

Physicians tell us that lunatics have often maladies of the ear, but none go further than Shakespeare's sweeping condemnation—

> O, then I see that madmen have no ears.
> *Romeo and Juliet*, iii. 3.

And when an evil deed has to be suggested, he ensures secrecy by saying—

> Hear me without thine ears.
> *King John*, iii. 3.

Macbeth mocks an apparition that calls him thrice by name, by replying—

> Had I three ears, I'ld hear thee.
> *Macbeth*, iv. 1.

The effect of music on the ear is shown in the well-known passage, which we give with the modern emendation of "south" (*i.e.* the south wind) for "sound"—

> That strain again! it had a dying fall:
> O, it came o'er my ear like the sweet south,
> That breathes upon a bank of violets,
> Stealing and giving odour.
> *Twelfth Night*, i. 1.

Shakespeare, the king of quips and cranks, could say—

> A jest's prosperity lies in the ear
> Of him that hears it, never in the tongue
> Of him that makes it.
> *Love's Labour's Lost*, v. 2.

# IN LITERATURE & SCIENCE

He also uses the old English verb, to *ear*, meaning to *plough*—

> He that *ears* my land spares my team.
> *All's Well*, i. 3.

Milton often uses phrases about the ear that are very impressive, as to hearing attentively or with pleasure, but he did not describe it. His lines about the voice of the Lady in *Comus* may be quoted—

> I was all ear,
> And took in strains that might create a soul
> Under the ribs of death.
> Milton's *Comus*, line 560.

Keats, who was considered to lay too much stress upon the delights of the senses, nevertheless exalts the powers of the imaginative ear above those of the physical ear. His exquisite lines upon the flute-players pictured on a Grecian urn express this in the very spirit of poetry—

> Heard melodies are sweet, but those unheard
>     Are sweeter; therefore, ye soft pipes, play on;
> Not to the sensual ear, but, more endear'd,
>     Pipe to the spirit ditties of no tone.
> Keats's *Ode on a Grecian Urn*.

The sonnet "What tongue can her perfections say?" in which Sir Philip Sidney praises the ear at the expense of the ear-ring, from which he draws a brilliant simile whilst, in fact, denouncing the custom, contains the following lines :—

> The tip no jewel needs to wear,
> The tip is jewel of the ear.

If we turn to prose, perhaps the best and most eloquent nomenclature that ever clattered about the ears will be found in Charles Lamb's *A Chapter on Ears*. It begins abruptly enough—

> I have no ear——
> Mistake me not, reader, nor imagine that I am by nature destitute of those exterior twin appendages, hanging ornaments, and (architecturally speaking) handsome volutes to the human capital. Better my mother had never borne me. I am, I think, rather delicately than copiously provided with those conduits; and I feel no disposition to envy the mule for his plenty, or the mole for her exactness, in those ingenious labyrinthine inlets—those indispensable side-intelligencers. . . . I even think that, sentimentally, I am disposed to harmony. But organically I am incapable of a tune.—Charles Lamb, *A Chapter on Ears*.

Science has an interesting word to say about the outer ear, although it is generally

# IN LITERATURE & SCIENCE 215

badly drawn in scientific illustrations. The chief interest to scientific men seems to lie in the inner ear, which is so carefully drawn in their books and contains such beautiful curves that we trust it is more accurately delineated than the outer ear. Nevertheless, each inner ear probably varies in consonance with the outer ear, and indeed the nerves are said to vary in number and size according to the powers of hearing. In my account of ears of musicians in opposition to ears that incline to deafness in early life, I drew attention to the wide, square orifice as betokening strong and acute natural powers of hearing. Probably the auditory nerves are stronger and more numerous in the inner ear that goes with an outer ear of this shape. It would be worth while to take extra care of children's ears where the narrow shapes are found, so as not to wear out the auditory nerves too soon, just as delicacy of eyesight is tended. Hitherto we have had no warning betimes, until the mischief is done or at least begun.

The outer ear, called the *auricle*, varies in shape for the purpose of catching the sounds at the best angle for reflecting them into the auditory canal. And when the waves have got into that curving canal, they splash backwards and forwards as it were, so that by the time they reach the drum of the ear they strike it straight on its membrane, giving the best volume of sound possible. Therefore it matters very much, even in the eyes of science, what shape the outer ear may be, because upon that depends the size and power of the waves of sound at the outset of their journey to the inner ear. The easiest way to make a somewhat deaf person hear what is said without an ear-trumpet, is for the speaker to be about ten inches off and to direct the voice straight into the orifice in a level stream of sound and with a very distinct articulation of the words. The ear should be kept exactly in the same place, the face of the listener being in profile. With scarcely the least raising of the voice, by observing these simple precautions to set the waves of sound going in a right direction for a concentrated effort to reach the dulled tympanum, every word can be distinctly heard. Deaf people often complain of being shouted at, for shouting bangs the waves of sound about like a storm hurtling the sea, confusing the words and the powers of attention. The advantage of being about ten inches off, is that the stream of sound from the voice has space to form into a compact column, giving more force with less effort to the speaker.

The following verses give a popular account of the mechanism of the human ear, including the scientific terms. It appeared first in the *Illustrated London News*, vol. xx. January 17, 1852, and was reprinted in *Notes and Queries*, April 29, 1871, from which we quote it entire :—

### THE PHILOSOPHER AND HER FATHER

A sound came booming through the air,
 "What is that sound?" quoth I.
My blue-eyed pet, with golden hair,
 Made answer, presently,
"Papa, you know it very well—
That sound—it is Saint Pancras' Bell."

My own *Louise*, put down the cat,
 And come and stand by me ·
I'm sad to hear you talk like that,
 Where's your philosophy?
That sound—attend to what I tell—
That sound was *not* Saint Pancras' Bell.

# THE HUMAN EAR

Sound is the name the sage selects
    For the concluding term
Of a long series of effects
    Of which the blow's the germ.
The following brief analysis
Shows the interpolations, Miss.

The blow, which when the clapper slips
    Falls on your friend the Bell,
Changes its circle to ellipse
    (A word you'd better spell).
And then comes elasticity,
Restoring what it used to be.

Nay, making it a little more,
    The circle shifts about
As much as it shrunk in before
    The Bell, you see, swells out;
And so a new ellipse is made
(You're not attending, I'm afraid).

This change of form disturbs the air,
    Which in its turn behaves
In like elastic fashion there,
    Creating waves on waves;
Which press each other outward, dear,
Until the outmost finds your ear.

Within that ear the surgeons find
    A *tympanum* or drum,
Which has a little bone behind,—
    *Malleus*, it's called by some;
But those not proud of Latin Grammar
Humbly translate it as the hammer.

The wave's vibrations this transmits
    On to the *incus* bone
(*Incus* means anvil, which it hits),
    And this transfers the tone
To the small *os orbiculare*,
The tiniest bone that people carry.

The *stapes* next—the name recalls
    A stirrup's form, my daughter—
Joins three half-circular canals,
    Each fill'd with limpid water;
Their curious lining, you'll observe,
Made of the auditory nerve.

This vibrates next—and then we find
    The mystic work is crown'd;
For then my daughter's gentle Mind
    First recognises sound.
See what a host of causes swell
To make up what you call "the Bell.

Awhile she paused, my bright *L*ouise,
    And pondered on the case;
Then, settling that he meant to tease,
    She slapped her father's face.
"You bad old man, to sit and tell
Such gibbergosh about a Bell!"

# INDEX

*Ante-helix*, described, 10
*Ante-tragus*, described, 9; varies greatly in outline, 31; well set back, 40
Apollonides, ears of the Lydian, 5. *See* XENOPHON.
Aristotle, supposititious work on *Physiognomy*, including a passage on the ears, 44; this is criticised by Pliny, 48

Bertillon, M. Alphonse, on the ears of criminals, 6; assumes ears do not alter from infancy, 124, *note*
Bristol Museum, Roman votive ears in the, 53; the votive ears described, 104
British and foreign ears, 2

Classification of ears, the system explained, 27
*Coalescing* forms of the Five Divisions, *explained*, 13; how to classify, 27; where they usually occur, 39

Divisions of the *helix*, the Five, 14; the four French ones, *ib.* (*see* Bertillon); the proportions of the Five, 15; the places of the Five, 17; the Five Divisions, separate and combined, arranged for the classification of ears, 27; notes on varieties of their form, 35; Division (4) too "high," like a "high" cheek-bone, 38; physiognomy of the Five Divisions, 131; method of using the Five Divisions, to catalogue certain qualities, 135. *See* HELIX.

Ear, the human, as a means of identification, 1; its outer parts, *described*, 3; cannot be altered at will, 5; alters in shape during childhood, 6; how to take a nature-print of the, 11; the Five Divisions of the *helix*, 15; the law of the shape and proportion of the, 16; the eight squares, applied to a nature print, 17; is the same size as the nose, 19; the outline of the shape agrees with the jaw in profile, 22; the perfect form for art, 23; the perfect form for science, 24; table of percentages of divisions of the *helix* in 100 ears, 26; classification, *explained*, 27; how to describe an ear for identification, 33; astrology was never applied to "govern" the ear, 43; supposititious work by Aristotle on the ear, 44; Pliny on the ear, 46; tingling of the ears, 49; mediæval writers on, 54; "dog-eared" men, 67; Ghirardelli's ten types, 69; Lavater on the ear, 86; the physiognomy of the ear, *ex-*

*plained*, 123; how the ears of a child alter, 124; the ear is full grown at fourteen, 129; no "fortune-telling" in the ear, 131; the probable manner in which the native qualities will be used is indicated in the ear, 135; the *right* ear differs often from the *left* ear, 136; the *nodule*, 143; ear of the musical composer, 144; ear with a "neck," 145; wide top, *ib.*; the straight top, 148; ear of a traveller, *ib.*; ear of an artist, 149; ear of an artist and traveller, 150; ear of an American, showing there is no indication of nationality in ears, 153; ear of a naval nurse, 154; ear of a zoologist, 155; ear of a philologist, 156; ear of the editor of the *English Dictionary*, 157; ears with peculiar orifices, 159; ear of a scientific discoverer, 160; ear with the Five Divisions in typical forms, 161; ear of a novelist (Charles Dickens), 163; heredity in the ear, 166; one orifice is often copied from the father and the other from the mother, 169; as the lobe follows the shape of the chin, this part of the ear is but little reproduced in the children, 173; great difference in the ears of twin-sisters, *explained*, 179; the ear in sculpture and painting, 180; Assyrian ear, *ib.*; Egyptian ear, 182; Greek ear, 186; ear of Julius Cæsar, 187; ear of Hermes, by Praxiteles, 188; defective modern sculpture of ear, 189; the ear is difficult to draw, 190; the extreme accuracy of the ear painted by Raffaelle, 191; truncated ear in portraits with wigs, 192; votive offerings of ear, ancient, 194; ditto, modern, 195; ear-rings, 198; ear-lore, 204; the ear-dove, 207; the ear-wort, *ib.*; charade on the ear-wig, 208; the folk-lore "Fine-ear," 209; the ear in literature, *e.g.* in poetry and prose, 210; on the scientific drawings of the ear (inner and outer), 215; poem on the mechanism of the human ear, 217

Ear-prints, taken direct from nature, 6; two sets taken from a child, showing alterations in growth and shape, 7; how to take, 11; their advantage over drawings in preserving the relative sizes of ears, 32; their appearance, *ib.*; ear-prints are reversed (like a negative of a photograph), 33; a *left* ear-print described in squares, *ib.*

Ears of (described)—
Augustus Cæsar, 68; Sir John Stainer, Kt., Mus. Doc., 144; Mozart, *ib.*; Mrs. Theodore Bent, 148; Miss Annette Elias, 149; Mr. A. Henry Savage Landor, 150; Miss E. Keogh, 155; Miss F. Buchanan, 156; Miss E. E. Wardale, Ph.D. of Zurich, *ib.*; Dr. J. A. H. Murray, 158; Sir John Burdon Sanderson, Baronet, M.D., F.R.S., etc., 161; Prof. A. H. Sayce, 162; Charles Dickens, 163; Lady Stainer, 168; Mr. J. F. R. Stainer, *ib.*; Dr. Richard Garnett, 175; Mrs. R. Garnett, *ib.*; Mrs. Guy Hall, *ib.*; Dr. Griffiths, 178; Mrs. Griffiths, *ib.*; Miss Griffiths, *ib.*; Mr. Cyril Griffiths, *ib.*; the ancient Assyrians and their captives, 180; the ancient Egyptians, 182; Tahutmes IV., 184; Amenophis III., 185; Rameses II., *ib.*; Pai, *ib.*; the ancient Greeks and Romans, *ib.*; Cicero, 186; Marcus Aurelius, *ib.*; Julius Cæsar,

# INDEX

187; Praxiteles, 188. *See also* LIST OF ILLUSTRATIONS.

England, French system of identification of criminals employed in, 14

Evelyn, his acute hearing, 46. *See Evelyn's Diary.*

"Five Divisions," the. *See* DIVISIONS OF THE HELIX.

French system of identification, used for criminals, *explained*, 14; how it differs from the new system in this book for "non-criminals," *ib. See* IDENTIFICATION.

Galton, Mr. Francis, on *finger-tip* identification, 4; his materials for printing from nature can be used for ear-prints, 12

Glass, piece of, with eight squares marked on it, 17; to be laid on an ear-print, *ib.*

Headings, fourteen, used for the classification of ears, 27; five subdivisions of each, 28

*Helix*, described, 7; varies in each ear, 9; the Five Divisions, 15; may be jagged in shape, 16; the outline follows the outline of the jaw in profile, 22; how the jags are formed, 41; alters during childhood, 124; sometimes nearly absent, 143. *See* DIVISIONS, PINNA.

Heredity, as shown in ears, 166; the Five Divisions are recombined in the children, not directly reproduced, *ib.*; the orifice, 169; the knot, 170; the top of the *pinna*, 171; the lobe, 173; Division (4) in relation to, 178

Hollow *or* orifice of the ear, how to estimate its relative size, 18; is rarely drawn or carved accurately, *ib.*; the square form, 144; the narrow form presages deafness, *ib.*; the pear-shape, 157

*Human Ear, as a Means of Identification*: paper by the author, read at the meeting of the British Association, Bristol (1898), 25

Identification, the human ear as a means of, 1; the sources of *data* for, 2; ears have been used hitherto for identification of criminals only, and those of the "non-criminals" have been overlooked, 8; the French system of identification for criminals, 14; the head of the Identification Office in England, *ib.*; the new system of identification by the ear for "non-criminals," 15; the fully-developed form of ear is rare, and is in itself a remarkable case for identification, 25; the parts of the ear chiefly distinctive for purposes of identification, 26; a trained eye is required to judge identification by the ear, 34; the new system of physiognomy of the ear is based upon observation and scientific forms of identification, 165

*Indent* in *helix*, described, 16; should be noted on the margin of the ear-print, 32; example of, 152

*Knot*, described (Darwin's "nodule"), 10; rare and valuable for identification, 31; not often found in Division (1), 35; seems to run in families, 41; appears to go with a power of specialising, 142

Lavater, Johann Caspar (1741-1801), could not account for pointed ears, 35; admired a squareness in the upper half of the ear, 37 ; sometimes drew the *helix* with a straight upright outline, 39 ; his life, 86 ; his book on Physiognomy, 87 ; his instinct for judging character, 89 ; his first account of ears, 91 ; said there was a "special physiognomy of the ears," *ib.* ; five ears, described, 92 ; his second account of ears, 94 ; twenty-one ears, described, 96 ; his third account of ears, 105 ; nine ears, described, 107 ; his fourth account of ears, 111; twelve ears, described, 113 ; his fifth account of ears, 120 ; three ears, described, 121. *See* also p. 92, *ante*.

Lobe, described, 9 ; differs constantly from its fellow, *ib.* ; its size varies according to the size of the chin in profile, 22 ; superstitions regarding, 40 ; is fully developed earlier than the chin, 173 ; how it varies in indirect reproduction from the parents' forms, *ib.* ; when it differs very much from the other 174

Mediæval writers on ears: Porphyry, 56 ; Averroes, *ib.* ; Albertus Magnus, *ib.* ; Pietro d'Abano (called *Conciliator*), 58; Bartolomeo Cocles, *pseudonym* Andrea Corvo, 59 ; Jean Taisnier, 60 ; Indagine, *ib.* ; Jean Belot, 61 ; Honoratus Niquetius or Nicquet, *ib.* ; Pomponius Gauricus, 63 ; Giovanni Ingegneri, *ib.* ; Giovanni Battista della Porta, 65 ; della Porta quotes Aristotle, Pliny, Galen, Polemon, Adamantio, Conciliator, Losso, Suetonius, Columella, and Meletius, 66 ; Cornelio Ghirardelli, 68 ; Dominico de Rubeis, 82

Moles, formerly counted as proofs of identity, 4 ; warts often mistaken for moles, *ib*.

Mozart, resemblance between the ears of Sir John Stainer, Kt., Mus.Doc., to those of, 143

Nature-prints of the ear, 6 ; nature-prints of a child's ears at different ages, 7 ; nature-prints should be taken of every adult, as a safeguard of identity, 8 ; how to take nature-prints of the ear, 11. *See* EAR-PRINTS.

Nemesis, resides behind the right ear, 47. *See* PLINY ON EARS.

"Nodule," Darwin's. *See* KNOT.

"Non-criminal" classes, greater variety among their ears, *explained*, 2 ; are photographed in childhood, showing growth of *pinna* (*q.v.*), 7 ; the value of ears for identification has been overlooked amongst the non-criminal classes, 8 ; the non-criminal classes appear to have more fully developed ears than the criminal, *explained*, 15 ; the new system of physiognomy in this book, though invented for the non-criminal classes, can also be applied to the criminal, 165. *See* BERTILLON, IDENTIFICATION, PHYSIOGNOMY.

Orifice of the ear. *See* HOLLOW OF THE EAR.

Otomorphology, *lit.* the science of the shape of the ear, 42

Physiognomy, supposititious work by Aristotle on, 44 ; mediæval writers on, 55 ; della Porta, Ghirardelli, and Rubeis on, 65 ; Lavater on, 87 ; a new physiognomy of the ear, 123 ; although this book deals only with the "non-criminal" classes, this new physiognomy can also be applied to criminals' ears, 165

# INDEX

*Pinna*, described, 3 ; the right *pinna* is generally the largest, 9 ; its shape (not its size) gives quick hearing, 11 ; its slope follows the slope of the nose, 21 ; its extension backwards "squares" the form at top or bottom, 23 ; neck to the *pinna*, 145

Pitt-Rivers Collection at the University Museum, Oxford, Roman votive ears in the, 53 ; ancient and modern votive ears in, described, 104

Pliny, on ears, 46 ; mentions tingling of the ears, 49 ; his remedies for diseases of the ears, 50. *See* PLINY'S *Natural History.*

Table of percentages of divisions of the *helix* in a hundred selected ears, 26

*Top* of the ear, described, 9 ; straight, 148 ; very pointed, *ib. See* HEREDITY.

*Tragus*, described, 9

Twins, easily distinguished by the ear, 8

Warts, often mistaken for moles (*q.v.*), 4

Xenophon, identification of Apollonides as being a Lydian by his pierced ears (*Anab.* Bk. III. ch. i. § 31), 5

THE END

*Demy 8vo, Cloth.  Price* **7s. 6d.** *net.  With 33 Illustrations.*

# THE GRAMMAR OF SCIENCE

*Second Edition, thoroughly Revised and much Enlarged*

By KARL PEARSON, M.A., F.R.S.

PROFESSOR OF APPLIED MATHEMATICS AND MECHANICS IN UNIVERSITY COLLEGE, LONDON

Contains two entirely New Chapters on Natural Selection and Heredity, embracing a popular account of Prof. Pearson's own more recent work in this direction.

### SCOPE OF THE BOOK.

This work attempts to give a philosophical basis to the fundamental principles of modern science. It assumes no special mathematical or biological training on the part of the reader, but endeavours to lay before the man with average education an intelligible account of what science professes to achieve and of what it does not. The first four chapters define the material and lay down the principles of all scientific reasoning; they explain the scope, methods, and hopes of science and its relation to our theory of life. The following four chapters discuss the axioms and principles of physical science, and endeavour to give a rational view of mechanism which is not open to the criticisms raised against it by Balfour, Ward, and other recent metaphysical writers. The next three chapters deal with the science of organic forms, discussing the principal factors of evolution and endeavouring to give them exact quantitative definition. The two chapters on evolution place before the reader the present position of the Darwinian theory, at the same time indicating the futility of recent reactionary attacks. The final chapter deals with the classification of the sciences, and gives a bird's-eye view of the fields wherein the specialist alone can work.

### PRESS NOTICES.

" Not the least interesting part of this powerful book is the discussion of the effect on the mind of a true scientific education, which enables a man or woman to form judgments freed from individual bias. . . . We recommend all readers, and especially scientists, metaphysicians, theologians, and last, but not least, the writers of scientific text-books, to read and digest this well-written, clearly reasoned description of what science and scientific method is."—*Pall Mall Gazette.*

" . . . We have been again and again impressed in examining *The Grammar* with the remarkable lucidity of Prof. Pearson's explanations. '—*Knowledge.*

" It is still a grammar in that it deals with the foundations of science, but a far more ambitious title might have been given to so comprehensive a work."—*The Bookman.*

---

A. & C. BLACK, SOHO SQUARE, LONDON.

BOOKS ON HEALTH.

By GEORGE S. KEITH, M.D., LL.D., F.R.C.P.E.

*Crown 8vo, Cloth. Price* **2s. 6d.** *each.*

## PLEA FOR A SIMPLER LIFE

*Sixth Edition.*

"It is the old exhortation, plain living and high thinking. But it is more, it shows the way to reach it. It is indeed a most earnest yet scientific exposition of the evil we do to our bodies and souls and spirits by mixed dishes and medicines. If we would follow Dr. Keith's advice and take his prescriptions, we should have less dyspepsia and less atheism amongst us, less need for doctors of medicine and less need for doctors of divinity."—*Expository Times.*

'There is no doubt whatever that the book is full of wise counsel."—*Edinburgh Medical Journal.*

---

## FADS OF AN OLD PHYSICIAN

*Third Edition.*

"Some very excellent hints of the best methods for the cultivation of a sound mind in a sound body are to be gleaned, by the layman even, from Dr. Keith's little treatise, 'The Fads of an Old Physician' (Black). The booklet forms a sequel to the author's 'Plea for a Simpler Life,' which was received with well-deserved recognition by medical and lay authorities. Would that all fads had as much solid worth as those of an Old Physician."—*Pall Mall Gazette.*

"Dr. Keith's fads bear a remarkable resemblance to common sense, and the book throughout is eminently readable and interesting as well as instructive."—*Scotsman.*

---

## ON SANITARY AND OTHER MATTERS

This volume, as the title indicates, treats of subjects varying somewhat in kind, but all pointing out errors on sanitary or economic matters which seriously affect the well-being of the community, and which, but for the strong resisting power of conventionalism, might be easily remedied, and with much advantage to all. The last three papers go mainly to confirm what the author has already brought forward as to the evils resulting from the present system of over-feeding, over-stimulation, and drugging, both in health and disease.

---

A. & C. BLACK, SOHO SQUARE, LONDON.

Printed in Great Britain
by Amazon